Amplitude Equations for Stochastic Partial Differential Equations

INTERDISCIPLINARY MATHEMATICAL SCIENCES

Series Editor: Jinqiao Duan *(Illinois Inst. of Tech., USA)*

Editorial Board: Ludwig Arnold, Roberto Camassa, Peter Constantin, Charles Doering, Paul Fischer, Andrei V. Fursikov, Fred R. McMorris, Daniel Schertzer, Bjorn Schmalfuss, Xiangdong Ye, and Jerzy Zabczyk

---

Published

Vol. 1: Global Attractors of Nonautonomous Dissipative Dynamical Systems
David N. Cheban

Vol. 2: Stochastic Differential Equations: Theory and Applications
A Volume in Honor of Professor Boris L. Rozovskii
eds. Peter H. Baxendale & Sergey V. Lototsky

Vol. 3: Amplitude Equations for Stochastic Partial Differential Equations
Dirk Blömker

Vol. 4: Mathematical Theory of Adaptive Control
Vladimir G. Sragovich

Vol. 5: The Hilbert–Huang Transform and Its Applications
Norden E. Huang & Samuel S. P. Shen

Vol. 6: Meshfree Approximation Methods with MATLAB
Gregory E. Fasshauer

Interdisciplinary Mathematical Sciences – Vol. 3

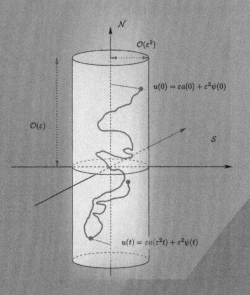

Amplitude Equations for Stochastic Partial Differential Equations

Dirk Blömker

RWTH Aachen, Germany

World Scientific

Published by

World Scientific Publishing Co. Pte. Ltd.

5 Toh Tuck Link, Singapore 596224

USA office: 27 Warren Street, Suite 401-402, Hackensack, NJ 07601

UK office: 57 Shelton Street, Covent Garden, London WC2H 9HE

British Library Cataloguing-in-Publication Data
A catalogue record for this book is available from the British Library.

AMPLITUDE EQUATIONS FOR STOCHASTIC PARTIAL DIFFERENTIAL EQUATIONS
Interdisciplinary Mathematical Sciences — Vol. 3

ISBN-13 978-981-270-637-9
ISBN-10 981-270-637-2

Printed in Singapore.

To Ulla, Ole Einar, and Stephanie.

Preface

This book is based on material obtained in my Habilitation at the RWTH Aachen. It presents a rigorous mathematical approach to the theory of amplitude equations for stochastic partial differential equations (SPDEs). This is a theory which is on a formal level well established and a valuable tool for physicists and applied mathematicians.

The goal is the approximation of SPDEs near a change of stability. It aims at providing details about the reduction of the dynamics of SPDEs to more simpler equations via amplitude or modulation equations. The key idea is the natural separation of time-scales present near a change of stability.

Main topics of this book are rigorous error estimates for the approximation by amplitude equations. These are well known for deterministic PDEs, and there is a large body of literature over the past about two decades. However there seems to be a lack of literature for stochastic equations, although without reliable error estimates at hand the theory in the stochastic case is successfully used. One celebrated examples is, for instance, the convective instability in Rayleigh-Bénard convection. This book is a first step in closing this gap.

The aims of the book are twofold. On one hand it presents a more elementary introduction to the subject highlighting the new tools necessary for stochastic equations. In a large part of the book not much expertise in stochastic partial differential equations is assumed. On the other hand it should provide a guideline to current research, by summarising results of recent research articles.

The author would like to thank his collaborators M. Hairer, S. Maier-Paape, G. Pavliotis, G. Schneider and, T. Wanner. Furthermore, J. Duan and World Scientific for their support.

Dirk Blömker

Contents

Preface   vii

1. Introduction   1

   1.1 Formal Derivation of Amplitude Equations   7
      1.1.1 Cubic Nonlinearities   8
      1.1.2 Other Types of Nonlinearities   10
      1.1.3 Quadratic Nonlinearities   11
      1.1.4 Large or Unbounded Domains   13
   1.2 General Structure of the Approach   17
      1.2.1 Meta Theorems   19
   1.3 Examples of Equations   21

2. Amplitude Equations on Bounded Domains   25

   2.1 Multiplicative Noise (Parameter Noise)   26
   2.2 Assumptions and Results — The Cubic Case   28
      2.2.1 Attractivity   32
      2.2.2 Residual   33
      2.2.3 Approximation   34
   2.3 A priori Estimates for u   37
   2.4 Results for Quadratic Nonlinearities   41
      2.4.1 Attractivity   44
      2.4.2 Residual   45
      2.4.3 Approximation   47
      2.4.4 Proofs   47
   2.5 Setting for Additive Noise (Thermal Noise)   55
      2.5.1 Assumptions   56
      2.5.2 Existence of Solutions   58
      2.5.3 Amplitude Equations — Main Results   59
         2.5.3.1 Attractivity   59

          2.5.3.2 Approximation 60
  2.6 Quadratic Nonlinearities 62

3. Applications — Some Examples   67

  3.1 Approximation of Invariant Measures 69
      3.1.1 The Results 72
  3.2 Pattern Formation Below Criticality 75
      3.2.1 Additive Noise 75
      3.2.2 Multiplicative Noise 79
  3.3 Approximative Centre Manifold 80
      3.3.1 Random Fixed Points 83
      3.3.2 Random Set Attractors 84

4. Amplitude Equations on Large Domains   89

  4.1 Introduction 89
  4.2 Setting 90
  4.3 Approximation of the Stochastic Convolution 93
      4.3.1 Noise 94
      4.3.2 Main Result 95
      4.3.3 Remarks 96
      4.3.4 The General Result 97
  4.4 Nonlinear Result 100

Appendix A   Basic Inequalities   103

A. Basic Inequalities   103

  A.1 Burkholder-Davis-Gundy Inequality 104
  A.2 Comparison Argument for ODEs 106

Appendix B   Bounds for SDEs   109

B. Bounds for SDEs   109

  B.1 Large Deviation Estimate 109
  B.2 Moment Inequalities 111
      B.2.1 Negative Moments 113

Bibliography   117

Index   125

Chapter 1

Introduction

Bifurcation theory for stochastic partial differential equations (SPDEs) is still not fully developed. In contrast to that stochastic ordinary differential equations (SDEs) especially with one-dimensional phase space are widely studied, and there are numerous results on bifurcation For instance, for one-dimensional SDEs a detailed classification of possible bifurcation scenarios is established. See for example [Ste00] or [CIS99]. But already with a two-dimensional phase space there are open problems.

An interesting feature of stochastic bifurcation theory is that there are several approaches describing bifurcations in SDEs, which sometimes yield completely different results. Numerous articles are devoted to the study of relations and differences of these concepts. Some examples of concepts are:

- Phenomenological bifurcation (or *P*-bifurcation), which characterises a bifurcation by using qualitative changes in the unique invariant measure for the corresponding Markov semigroup given by the law of stationary solutions.
- *D*-bifurcation (sometimes also called dynamical bifurcation), which is characterised by the changes in the set of random invariant measures for random dynamical systems.

For a *D*-bifurcation one could simply look at the change in the structure of the random attractor, which is a compact random set attracting all orbits or all bounded sets. See for instance [Arn98] for the theory of random dynamical systems and for examples of possible bifurcations (see Chapter 9 of [Arn98]). The connection between random attractors and invariant measures is for instance studied in [Sch99; Cra01]. Furthermore first steps are provided towards an abstract bifurcation theory in the spirit of catastrophe-theory for dynamical systems. See for example [DW06a; DW05] or [DTZ05b; DTZ05a].

Phenomenological Bifurcation

If we consider phenomenological bifurcation for equations with additive noise, then it is usually easy to derive the uniqueness of invariant measures under very mild non-degeneracy conditions on the noise. See [DPZ96] for a textbook and for examples of SPDEs with mild degeneracy conditions see [EM01] or [EH01] and the references therein. Thus on the level of Markov-processes the long-time evolution is well described by this invariant measure.

Furthermore, one can study the invariant measure as the solution of the well known Kolmogorov equation, which is extensively studied (see for example [Cer01] or [DPZ02] or [RS04; RS06; KRZD99]. But in contrast to that there are hardly any examples describing qualitatively the structure of the invariant measures.

In Sections 3.1 and Theorem 4.4 we describe the results obtained in [BH04] and [BHP05]. These are two applications, where one actually can describe the structure of the invariant measure near a bifurcation. Hence, it is possible to detect phenomenological bifurcations in these models.

Random Attractors

There are many examples deriving the existence of random attractors for SPDEs. See for example [CF94; CDF97] for the definition of random attractors and for instance [FS96; FS99; DKS01] for some applications. But there are only a few results on the fine structure of the random attractors. For instance, upper bounds for the dimension are well known, but for many examples with additive noise, the random attractor is just a single random point.

For multiplicative noise the structure of a random attractor near a deterministic fixed point was studied in [CLR01]. It was shown that the random attractor actually undergoes a bifurcation, where the attractor changes from a single point to a random set. This was achieved by looking at the local dimension of the corresponding stable and unstable invariant manifolds in a small neighbourhood near the fixed point. For the general theory of invariant manifolds for SPDEs see [DLS03] or [MZZ07], but these results focus mainly on existence.

Random attractors for SDEs or SPDEs have been the topic of intense research. But when we look at the structure of a random attractor, it is just a single random point for many examples with additive noise. See for instance the celebrated work of Crauel and Flandoli [CF98] showing that for a one-dimensional SDE with additive noise there is no bifurcation on the level of random attractors. This was extended to monotone SPDEs in [CS04]. Similar results for non-monotone systems were derived by [Tea06b; Tea06a]. These results show that different concepts of bifurcation yield different results, as most of these examples do exhibit a phenomenological bifurcation. But the bifurcation can only be seen from the probability distribution or the dynamics of the random fixed point in time.

The problem of the attractor being a single random point is in general open. Even for simple two-dimensional phase space, there are only results for very small or very large noise or monotone SDEs.

Here we take a somewhat different approach. Instead of looking at the bifurcation of objects for time to ∞, we look at the typical transient behaviour of solutions of SPDEs. Here transient means possibly large, but still finite, time intervals. But even for SDEs the typical transient dynamics is not very well described in many examples. One exception being the work of Berglund and Gentz (see for instance [BG02a], [BG03], or the book [BG06]), where a detailed description of dynamics was derived for low-dimensional phase-spaces. Our results allow to carry some of these results immediately over to SPDEs. See for example [Blö03].

Freidlin-Wentzell Theory

We focus on equations with small noise in a neighbourhood of a change of stability. A similar but slightly different approach is the celebrated theory of Freidlin and Wentzell on large deviation effects. See [FW98] or Section 12 of [DPZ92]. One topic in this theory is to point out the effect of very small noise on the exit from domains of attraction of the deterministic model. Nevertheless, these effects occur only on time-scales, which are exponentially large in the noise strength, as in the small noise limit the solution follows the deterministic model for most of the time with very high probability. Thus, as the noise is induced for instance by very small thermal fluctuations, all these effects are obviously difficult to detect in experiments.

For Freidlin-Wentzell theory of SPDEs see [DPZ92; OV05]. For application to Cahn-Hilliard and Allen-Cahn see [Gaw06; Bra91; BGW07].

Our point of view is different from classical Freidlin-Wentzell Theory. We consider not only small noise, but also small parameters in the equation. This reflects the fact, that we are interested in experiments, where there is on one hand a very small source of noise (e.g. thermal fluctuations), which is usually described as the small noise limit of noise intensity to 0. But on the other hand one also tries to get the parameters sufficiently close to the change of stability, in order to see the effects of this bifurcation on the dynamics of the model, which is a second limit to 0.

One of our interest is to have a coupling of these two limits. In the context of Freidlin-Wentzell theory, this was only studied by using first the small noise limit, and then later on the limit of the parameter to 0 in the action functional, but not both limits at the same time. See for example [BDP05; OKRVE06; RK06].

Multi-Scale Analysis

The approach presented here relies on the fact that near a change of stability the dynamics of (stochastic) PDEs is driven by some dominant modes. These are exactly the eigenfunctions of the linearised operator, with eigenvalues close to 0. All other modes are subject to strong linear damping. Amplitude equations describe these essential dynamics. This is well known in the physics community and rigorously established for the deterministic PDEs. See for example [CE90; KSM92; Sch94; MS95] or for a recent review on bifurcation theory on unbounded domains [Sch01]. For PDEs on unbounded domains these amplitude equations are also known as modulation equations.

On a formal level, as already mentioned, there are numerous results using non-rigorous multiple scale analysis to derive reduced descriptions for the dynamics. See for example [CH93] or [Hak83] both containing plenty of such formal approximations. An interesting example is [Kus03] for highly oscillating solutions of an SDE.

First rigorous results are possible do to the theorem of Kurtz [Kur73] but, based on abstract semi-group theory for generators of Markov-diffusions, it provides no error bound, just the weak convergence of the approximation. Furthermore, it is a result for SDEs and not for SPDEs. See for example [WSPS04; MTVE01] for applications.

Pattern Formation Below Threshold of Stability

If the distance from the change of stability is sufficiently small, then the influence of small noise is detected in experiments. See for example the work of Ahlers, Rehberg et.al. on pattern formation below threshold of instability in convection problems. In Bénard-type models for electro-convection of liquid crystals [SA02; SR94] and in Rayleigh-Bénard convection for fluids [OdZSA04; OA03], it was verified experimentally, that near a change of stability there is a significant impact of thermal noise on the dynamics, leading to the formation of pattern in otherwise deterministically stable equations. This was long conjectured (see for example [CH93; HS92] and the references therein). The main difficulty of the experiment was to stabilise the control parameters (for example the temperature in Rayleigh-Bénard convection) to the precision of the noise strength, which is extremely small in case of thermal fluctuations in fluids. This is the main reason, why the effect was seen in electro-convection first.

The non-rigorous explanation of the experimentally verified effects relies on a formal expansion of the solution in the noise strength and separation between slow and fast components, in order to derive an effective equation for the amplitudes of the dominating pattern. This is also known as multi-scale expansion. We present results using this type of formal calculation in Section 1.1.

Amplitude Equation

The main object of this book is to rigorously establish estimates for the approximation of SPDEs via amplitude equations, and to answer questions on how noise influences the dynamics in systems near a change of stability. This has a lot of interesting applications, like pattern formation below criticality (see Section 3.2) or the structure of invariant measures for SPDEs (see Section 3.1), which gives insight into phenomenological bifurcations in SPDEs.

On bounded domains for SPDEs the approximation via amplitude equations was first rigorously verified in [BMPS01] for a simple Swift-Hohenberg model, and later on extended in [Blö03; Blö05a; BH04]. We outline typical results in Section

1.2 in a non-technical way. Here the amplitude equation for the dominant modes is given by an SDE. A typical example is

$$\partial_T \mathcal{A} = \nu \mathcal{A} - c|\mathcal{A}|^2 \mathcal{A} + \dot{\beta} \,,$$

where $\mathcal{A}(T) \in \mathbb{R}^n$, $\nu, c \in \mathbb{R}$, and $\dot{\beta}$ is noise in $\mathbb{R}^n$.

In Chapter 2 we review these results in detail and present the theory applied to a simple model with multiplicative noise, also known as parameter noise. In that model rigorous results were not considered before.

While all results for amplitude equations are mainly limited to transient behaviour, it was in [BH04] also possible to approximate the long-time behaviour in terms of the structure of invariant measures for the corresponding Markov-semigroup. See also Section 3.1 or [BH05].

The case of unbounded or just very large domains is significantly different. The amplitudes of the dominant modes are subject to a long-range modulation in space, and hence not given by an SDE, but an SPDE instead. The celebrated example is an SPDE of Ginzburg-Landau type:

$$\partial_T A = \partial_X^2 A + \nu A - c|A|^2 A + \xi \,, \tag{1.1}$$

where $A(T, X) \in \mathbb{C}$ and ξ is space-time white noise.

The case of large, but still bounded, domains is discussed in [BHP05]. See also [MSZ00] for the deterministic equation. In both cases the domain is bounded, but it scales with respect to the distance from bifurcation. A review of the stochastic result can be found in Chapter 4. As already mentioned before, there is also a very large literature for deterministic equations on unbounded domains, but this seems to be out of reach for SPDEs.

In all of these articles for the stochastic case, we consider noise that is on one hand sufficiently small, as given from the experiment, but on the other hand sufficiently large compared to the distance from the bifurcation. As discussed before, the viewpoint of the experiment is that we try to adjust the bifurcation parameter sufficiently close to the bifurcation, in order to see both stochastic effects and the small linear stability or instability in the amplitude equation. Of course, we could look at different scalings between these two small quantities, loosing either the noise or the linear (in)stability in the approximation. Depending on the type of equation the scaling of the coupling between the noise strength and the distance from bifurcation may change, in order to get an interesting stochastic amplitude equation.

The main difference between small and large domains is the existence of a large spectral gap of order $\mathcal{O}(1)$ in the linearised operator of the PDE. On bounded domains, we have a finite number $e = (e_1, \dots, e_n)$ of modes (or eigenfunctions) such that the corresponding eigenvalues change sign at the change of stability. If we are close to the bifurcation, all other eigenvalues are negative and sufficiently far away from 0. Formal arguments (see for instance Section 1.1.1) show, that the amplitudes $\mathcal{A} \in \mathbb{R}^n$ of the dominating modes are given by the so called amplitude equation, while the solution u of the SPDE is well approximated by

$$u(t,x) = \varepsilon \mathcal{A}(\varepsilon^2 t) \cdot e(x) + \mathcal{O}(\varepsilon^2) \, ,$$

where ε^2 is the typical scale for the distance from bifurcation.

On unbounded or just very large domains this picture changes completely. Even very close to the bifurcation a large number of modes are near or already above the threshold of stability, but still small. In this case the amplitude $\mathcal{A}$ is also a function in x that is concentrated in Fourier space near the dominant modes. Hence, $\mathcal{A}$ is subject to slow modulations in space, taking into account the large number of weakly (un)stable modes. In Section 1.1.4 we see that the solution u is in this case given by

$$u(t,x) = \varepsilon \mathcal{A}(\varepsilon^2 t, \varepsilon x) \cdot e(x) + \mathcal{O}(\varepsilon^2)$$

and $\mathcal{A}$ fulfils a (stochastic) PDE, which is called *amplitude or modulation equation*. The typical example being a stochastic Ginzburg-Landau equation, where for instance $e = (\sin, \cos)$, and we can thus write

$$u(t,x) = \varepsilon A(\varepsilon^2 t, \varepsilon x) e^{ix} + c.c. + \mathcal{O}(\varepsilon^2)$$

where the complex amplitude $A(T,X) \in \mathbb{C}$ solves a Ginzburg-Landau equation similar to (1.1).

A key point is that the amplitude equation for $\mathcal{A}$ is in a certain sense independent of ε. Although the paths of the process $\{\mathcal{A}(T)\}_{T \geq 0}$ might depend explicitly on ε, this dependence disappears on the level of probability measures due to the scaling properties of the noise. This is explained in more detail in Remark 1.2. Thus any result for the amplitude equation immediately carries over to a rescaled statement for the original equation, provided that $\varepsilon > 0$ is small. This is for instance important in the results on pattern formation. See Section 3.2. One could also refer to this fact as a type of normal form theory, as the amplitude equation for many examples will always be of the same Ginzburg-Landau type.

In the remainder of the introduction, we first present results on the formal derivation of the amplitude equation in Section 1.1. Cubic nonlinearities are treated in Section 1.1.1, where we also take higher order corrections into account. Quadratic nonlinearities exhibit special properties, as in many examples dominant modes are not mapped to dominant modes by the nonlinearity. This will be addressed in Section 1.1.3, while in Section 1.1.4 the effect of large domains is studied. Section 1.2 presents the general method of proof, which is similar for all cases, and states typical results. While the final Section 1.3 of the introduction presents some examples of SPDEs, where the general theory applies.

1.1   Formal Derivation of Amplitude Equations

In this section, we discuss the formal derivation of amplitude equations and higher order corrections. Therefore, we use multiple scale analysis to reduce the equation to the essential dynamics, which involves the expansion of all terms in a small parameter. This is well known for many examples. Here we present results described in more detail for quadratic nonlinearities in [Blö05a] and for cubic nonlinearities in [BH04]. For large domains we summarise results of [BHP05] in Section 1.1.4.

Let us consider parabolic semilinear SPDEs or systems of SPDEs perturbed by additive forcing near a change of stability. Let us suppose, that the noise is of the order of the distance from the bifurcation. The use of additive noise is mainly for simplicity of presentation, and it is not very restrictive. We comment on multiplicative noise later in several occasions in Chapter 2. A large body of the research papers are on additive noise, which we will summarise later. In this book simple multiplicative noise is used to present a self-contained introduction to the topic.

The general prototype is an equation of the type

$$\partial_t u = Lu + \varepsilon^2 Au + \mathcal{F}(u) + \varepsilon^2 \xi \,, \tag{1.2}$$

where

- L is a symmetric non-positive differential operator
  (e.g. $1 + \partial_x^2)^2$) with non-zero finite dimensional kernel (or null-space),
- $\varepsilon^2 Au$ is a small (linear) deterministic perturbation,
- $\mathcal{F}$ is some nonlinearity, for instance a stable cubic like $-u^3$.
- $\xi = \xi(t,x)$ is a Gaussian noise in space and time

We later give examples of the noise, which is taken to be white in time and can be either white or coloured in space. To be more precise, suppose that ξ is a generalised Gaussian process such that for mean and correlation

$$\mathbb{E}\xi(t,x) = 0 \quad \text{and} \quad \mathbb{E}\xi(t,x)\xi(s,y) = \delta(t-s)q(x-y) \,,$$

for some suitable spatial correlation function (or distribution) q. If q is the Delta-distribution δ, too, then we call ξ space-time white noise. In this case $\xi = \partial_t W$ is the generalised derivative of a cylindrical Wiener-process $\{W(t)\}_{t\geq 0}$ in a suitable Hilbert space. This means

- $W(0) = 0$
- The increments $W(t) - W(s)$ are independent for disjoint intervals (s,t).
- The path $t \mapsto W(t)$ is continuous with probability 1.
- $W(t)$ is normal with covariance operator $t \cdot Id$.

For general q we can always write

$$\xi = \partial_t QW$$

for some symmetric linear operator Q. We will state details later when necessary
For a detailed discussion see [DPZ92] or Section 2.5.

For the formal calculation in the bounded domain case, we rely mainly on
the scaling properties of the noise in time. To be more precise, we need that
$\varepsilon^{-1}\xi(\varepsilon^{-2}T, x)$ and $\xi(T, x)$ are for all $\varepsilon > 0$ versions of the same noise process, which
means that they are different ε-dependent processes, but their law is the same. This
is easy to verify on the level of correlation functions, using the scaling properties
of the Delta-distribution, or on a more rigorous level, one can use the the scaling
properties of W (cf. Assumption 1.1).

Remark 1.1   *For the rigorous results, we work with an integrated version of (1.2),
using mild solutions in some Hilbert space X. The precise definition of this is given
for instance in Proposition 2.1. Nevertheless, for the formal calculation we directly
work with the SPDE. This is usual for formal calculations in physics. The standard
mathematical way of writing the SPDE is*

$$du = (Lu + \varepsilon^2 Au + \mathcal{F}(u))dt + \varepsilon^2 dQW ,$$

where one uses Itô-differentials.

1.1.1   Cubic Nonlinearities

One interesting example of an equation with cubic nonlinearity is the Swift-
Hohenberg equation, which was first used as a toy model for the convective instability in the Rayleigh-Bénard problem (see [SH77] or Section 1.3).

On a formal level for the Swift-Hohenberg equation the derivation of the amplitude equation is well known, see for instance (4.31) or (5.11) in the comprehensive
review article [CH93] and references therein. The amplitude equation for (1.3) was
already treated in [BMPS01]. But here we follow the presentation from [BH04],
taking into account second order corrections.

The formal SPDE is

$$\partial_t u = -(1 + \Delta)^2 u + \varepsilon^2 \nu u - u^3 + \varepsilon^2 \partial_t QW . \tag{1.3}$$

It is obviously of the type of (1.2) with $L = -(1+\Delta)^2$, $A = \nu I$ for some $\nu \in [-1, 1]$,
and $\mathcal{F}(u) = -u^3$. We can for instance consider periodic boundary conditions on the
domain $[0, 2\pi l]^d$ for dimension $d \in \mathbb{N}$ and integer length $l > 0$. This is mainly for
convenience to ensure that the change of stability is exactly at $\nu = 0$. After slight
modifications we can also treat non-integer length $l > 0$ or non-squared domains.

For the formal derivation in this section we consider an equation of the type
(1.2) or (1.3) and assume:

Assumption 1.1   *Let $\{QW(t)\}_{t\geq 0}$ be a Q-Wiener process. This implies especially
that $\{W(t)\}_{t\geq 0}$ and $\{\varepsilon W(\varepsilon^{-2}t)\}_{t\geq 0}$ are in law the same process.*

Furthermore, let $\mathcal{F}$ be cubic (i.e. $\mathcal{F}(u) = \mathcal{F}(u, u, u)$ is trilinear).

Denote the kernel (or nullspace) of L by $\mathcal{N}$ and the orthogonal projection onto $\mathcal{N}$ by P_c. Define $P_s = I - P_c$.

Then we make the following ansatz:

$$u(t) = \varepsilon a(\varepsilon^2 t) + \varepsilon^2 b(\varepsilon^2 t) + \varepsilon^2 \psi(t) + \mathcal{O}(\varepsilon^3) \,, \tag{1.4}$$

with $a, b \in \mathcal{N}$ and $\psi \in \mathcal{S} := \mathcal{N}^\perp$ the orthogonal complement of $\mathcal{N}$ in X.

This ansatz is motivated by the fact that, due to the linear damping of order one in $\mathcal{S}$, the modes in $\mathcal{S}$ are expected to evolve on time scales of order one, whereas the modes in $\mathcal{N}$ are expected to evolve on much slower time scales of order ε^{-2}, as the linear operator is of order ε^2. This is mainly due to the fact that we have a well defined spectral gap of order $\mathcal{O}(1)$ between 0 and the first non-zero eigenvalue together with a small linear perturbation of order ε^2.

We do not use lower order terms, as we expect that small solutions stay small. Furthermore, using linear and nonlinear stability, it is possible to verify a priori estimates that rigorously verify that the typical scaling of a solution corresponds to the one prescribed by the ansatz (1.4). The statement is called the *attractivity* result (cf. Section 1.2).

Let us now come back to the formal derivation. Plugging the ansatz (1.4) into (1.2), rescaling to the *slow time-scale* $T = \varepsilon^2 t$ and expanding in orders of ε, we obtain by collecting all terms of order ε^3 in $\mathcal{N}$

$$\partial_T a(T) = A_c a(T) + \mathcal{F}_c(a(T)) + \partial_T \beta(T) \,. \tag{1.5}$$

Here,

$$\beta(T) = \varepsilon P_c Q W(\varepsilon^{-2} T) \,, \qquad T \geq 0$$

is a Wiener process in $\mathcal{N}$ with law independent of ε, due to the scaling properties of the Wiener process. We used

$$A_c = P_c A \quad \text{and} \quad \mathcal{F}_c = P_c \mathcal{F}$$

for short.

This approximating equation in (1.5) is called *amplitude equation*, as it can by rewritten to an SDE for the amplitudes of an expansion of a with respect to a basis in $\mathcal{N}$. Results like this well known for many examples in the physics and applied mathematics literature (for example [CH93, (4.31),(5.11)]). Moreover, there are numerous variants of this method. However, most of these results are non-rigorous approximations using this type of formal multi-scale analysis.

Remark 1.2 *One key point is that (1.5) is, at least in law, completely independent of the small parameter $\varepsilon > 0$, as the amplitude equation is in general ε-independent. Therefore, we do not use an index ε for a. Although the paths of the Brownian motion β in the amplitude equation (1.5) depend obviously on ε due to the rescaling in time, the law of β is independent of ε. This is due to the scaling properties of* a Wiener process, which state that the process $\{\varepsilon W(\varepsilon^{-2}t)\}_{t\geq 0}$ *is a version of the Wiener process* $\{W(t)\}_{t\geq 0}$ *for all* $\varepsilon > 0$. *Therefore, by the weak uniqueness the law of solutions of (1.5) is independent of* ε. *Thus we also neglect the dependence of* a *on* ε *in the notation, as it is only path-wise.*

Higher Order Corrections

We already verified, that the slow modes (or amplitudes of the dominant modes) decouple from the fast modes, at least approximatively to the first order in ε. We now see that this remains true for the second order corrections, too.

Collecting terms of order ε^2 in $\mathcal{S}$ yields for ψ the linear equation

$$\partial_t \psi(t) = L_s \psi(t) + P_s \xi(t) , \tag{1.6}$$

where we defined $L_s = P_s L$.

One can show furthermore that $b = 0$. Therefore, we first rescale ψ to the slow time-scale to obtain formally that in law

$$\psi(T\varepsilon^{-2}) = \varepsilon L_s^{-1} P_s \xi(T) + \text{``higher order terms''} .$$

Thus the term $\psi(T\varepsilon^{-2})$, when viewed on the slow time-scale, gives a contribution of order ε. Using this, we obtain for terms of order ε^4 in $\mathcal{N}$

$$\partial_T b = A_c b + 3\mathcal{F}_c(a, a, b) .$$

Since the initial condition for the equation is $u(0) = \varepsilon a(0) + \varepsilon^2 \psi(0)$ by the ansatz, one has $b(0) = 0$, and therefore b vanishes identically.

It is not easy to proceed to higher order corrections, as they do involve couplings of fast and slow modes, which will make the approximating equations much more complicated. Moreover, we formally need to justify quadratic functionals of the noise term, which is highly nontrivial. See [BH04] for a detailed discussion.

1.1.2 *Other Types of Nonlinearities*

Cubic nonlinearities are not very special, we can extend the simple idea of the previous section, using scaling and projection, to a lot of different types of nonlinearities. If we look at suitable scalings of the noise and the linear (in)stability we obtain in all cases interesting results. If we do not adapt the scaling, we either loose the linear instability or the noise in the amplitude equation.

Suppose for this section that $\mathcal{F}^{(n)}$ is some multi-linear nonlinearity, which is homogeneous of degree $n \in \mathbb{N}$ with $n \geq 2$ (i.e. for $\alpha > 0$, $\mathcal{F}^{(n)}(\alpha u) = \alpha^n \mathcal{F}^{(n)}(u)$). Then the noise strength in the SPDE (1.2) should be changed to $\sigma^2 = \varepsilon^{(n+1)/(n-1)}$ instead of ε^2. Thus the equation reads in the interesting scaling

$$\partial_t u = Lu + \varepsilon^2 Au + \mathcal{F}^{(n)}(u) + \varepsilon^{(n+1)/(n-1)}\xi . \tag{1.7}$$

Now with the ansatz

$$u(t) = \varepsilon^{2/(n-1)} a(\varepsilon^2 t) + \varepsilon^{(n+1)/(n-1)} \psi(t) + \mathcal{O}(\varepsilon^{2n/(n-1)}) \tag{1.8}$$

and a similar formal calculation as in the previous section, we derive the same type of amplitude equation. First collecting all terms of order $\varepsilon^{2n/(n-1)}$ in $\mathcal{N}$ yields

$$\partial_T a = P_c A a + P_c \mathcal{F}^{(n)}(a) + \partial_T \beta , \tag{1.9}$$

which now contains a nonlinearity which is homogeneous of degree n. The second order correction is exactly the same (cf. (1.6)) as in the cubic case, but now it contains all terms in $\mathcal{S}$ of order $\varepsilon^{(n+1)/(n-1)}$.

We will not focus on rigorous results for this type of equations, as they are very similar to the cubic case. After minor modifications one can easily transfer all results to the general case.

Remark 1.3 *Let us finally comment on the distance from the bifurcation, that is necessary in order to see the stochastic effects in the amplitude equation. The order of magnitude of the distance in terms of the noise strength σ^2 is*

$$\varepsilon^2 = \sigma^{4(n-1)/(n+1)} \approx \begin{cases} \sigma^4 & \text{for } n \to \infty \\ 1 & \text{for } n \to 1 \end{cases} .$$

Thus, the higher the order of the nonlinearity, the nearer we have to go to the bifurcation, in order to see the influence of the noise on the dominant modes, at least on the time-scale under consideration. Let us finally change the point of view and express the noise in terms of $\varepsilon > 0$. Now

$$\sigma^2 = \varepsilon^{(n+1)/(n-1)} \approx \begin{cases} \varepsilon & \text{for } n \to \infty \\ 0 & \text{for } n \to 1 \end{cases} .$$

1.1.3 *Quadratic Nonlinearities*

An interesting feature of quadratic nonlinearities $B(u) = B(u, u)$ is that in many examples $P_c B(a) \equiv 0$ for all $a \in \mathcal{N}$. In this case, the ansatz (1.8) yields only the linearisation. See (1.9). This means that we still look at solutions that are too small to capture any of the nonlinear effects present in the equation. In order to obtain a nonlinear amplitude equation, we either consider larger noise, or we look at a parameter regime where we are nearer to the change of stability.

To illustrate this problem, we briefly discuss a one-dimensional Burgers' equation, which is given by

$$\partial_t u = \partial_x^2 u + \mu_\varepsilon u - u \partial_x u + \sigma_\varepsilon \xi .$$

Let ξ be space-time white noise for simplicity.

Example 1.1   For periodic boundary conditions on $[0, 2\pi]$ and $\mu_\varepsilon = \mathcal{O}(\varepsilon^2)$, we readily obtain for the space of dominant modes $\mathcal{N} = \text{span}\{1\}$. All other eigenfunctions of ∂_x^2 are spanned by $\sin(kx)$ and $\cos(kx)$ with $k \in \mathbb{N}$ and eigenvalues $-k^2 \leq -1 < 0$. But now for $B(u) = u\partial_x u$ we have $B(1) = 0$, obviously.

Example 1.2   Consider Dirichlet boundary conditions on $[0, \pi]$, then the linear instability arises for $\mu_\varepsilon = 1 + \mathcal{O}(\varepsilon^2)$. Furthermore, $\mathcal{N} = \text{span}\{\sin\}$, as all other eigenfunctions are spanned by $\sin(kx)$ with eigenvalues again bounded by -1. But now $B_c(\sin) = 0$, where we used again the short-hand notation $B_c = P_c B$ and $B_s = P_s B$.

Hence, in both examples with the ansatz (1.4) we only derive the linearisation as the amplitude equation.

There are numerous examples in the physics literature of equations with quadratic nonlinearities and the same property, as described above. One model is the growth of rough amorphous surfaces. See for example [BGR02] and the references therein. Another example is the Kuramoto-Sivashinsky equation, but the probably most important one is the Rayleigh-Bénard problem which is the paradigm of pattern formation in convection problems. All equations are described briefly in Section 1.3.

If we want to take into account nonlinear effects, then we have to look at the coupling of the slow dominant modes to the fast modes. In the following, we will follow the presentation of [Blö05a], in order to briefly comment on the formal derivation of the amplitude equation in this case.

Consider an equation of the type

$$\partial_t u = Lu + \varepsilon^2 Au + B(u, u) + \varepsilon^2 \xi \qquad (1.10)$$

with

- B a symmetric and bilinear operator with $B_c(a, a) = 0$ for $a \in \mathcal{N}$.
- L, A, and ξ as in (1.2)

We make the following ansatz, which is significantly different from (1.8):

$$u(t) = \varepsilon a(\varepsilon^2 t) + \varepsilon^2 \psi(t) + \mathcal{O}(\varepsilon^3),$$

with $a \in \mathcal{N}$ and $\psi \in \mathcal{S}$, in order to take into account both nonlinear and noise terms in the amplitude equation. Our new ansatz yields in lowest order in ε the following system of formal approximations. First of order $\mathcal{O}(\varepsilon^2)$ on the fast time-scale t in $\mathcal{S}$.

$$\partial_t \psi(t) = L_s \psi(t) + B_s(a(\varepsilon^2 t), a(\varepsilon^2 t)) + P_s \xi(t). \qquad (1.11)$$

Secondly of order ε^3 in $\mathcal{N}$ on the slow time-scale $T = \varepsilon^2 t$

$$\partial_T a(T) = A_c a(T) + 2B_c(a(T), \psi(\varepsilon^{-2} T)) + \partial_T \beta(T), \qquad (1.12)$$

where the rescaled projection of the Wiener process β was defined after (1.5).

These equations (1.11) and (1.12) are a coupled system of equations. On one hand there is a dominating equation (1.12) on a slow time-scale, which is similar to the amplitude equation (1.5). This is coupled to an equation (1.11) on the fast time-scale. Equations with a similar structure are rigorously treated in [BG03; BG06] for SDEs, or in [PS03; KPS04] where tracers in a fast moving random velocity field are considered. There is also a review [GKS04] and numerous other references.

The aim is now to get an effective equation for the slow component completely independent of the fast modes. First rescale (1.11) to the slow time-scale $T = \varepsilon^2 t$ by $\psi(t) = \Phi(\varepsilon^2 t)$. Thus

$$\varepsilon^2 \partial_T \Phi = L_s \Phi + B_s(a, a) + \varepsilon P_s \hat{\xi} \, ,$$

where $\hat{\xi}(T) = \varepsilon^{-1} \xi(\varepsilon^{-2} T)$ is a rescaled version of the noise ξ. Now neglect all terms, which are not $\mathcal{O}(1)$. This yields the stationary problem

$$0 = L_s \Phi + B_s(a, a) \, .$$

As L_s is invertible on $\mathcal{S}$, we get in lowest order of ε that

$$\Phi = -L_s^{-1} B_s(a, a) \, .$$

This together with (1.12) establishes a single approximation equation.

$$\partial_T a = A_c a - 2 B_c \Big(a, L_s^{-1} B_s(a, a) \Big) + \partial_T \beta \, , \tag{1.13}$$

Surprisingly, this equation involves a cubic nonlinearity, although the nonlinearity in the original equation was quadratic.

There is a whole zoo of other effects appearing, if we consider (1.10) with noise of strength ε, but not action on the dominant modes $\mathcal{N}$ directly. See [Rob03] for the first observations and [BHP06] for the first rigorous treatment.

1.1.4 *Large or Unbounded Domains*

For unbounded domains the results are very different. First of all, we do not have a spectral gap, and near the change of stability a whole band of eigenvalues gets unstable. The same effect already occurs, if we consider large domains, which are at least of the size $\mathcal{O}(\varepsilon^{-1})$. In Figure 1.1 we briefly sketch the eigenvalue curve $k \mapsto -P(-k)$ with the corresponding eigenvalues of the Swift-Hohenberg operator $-P(i\partial_x) = -(1 + \partial_x^2)^2$. For the deterministic PDE this somewhat intermediate step was already discussed in [MSZ00]. The stochastic case is treated in [BHP05], but we present a different formal derivation here. This is closer to usual physical reasoning, and more in the spirit of [KSM92].

Consider as an example a one-dimensional version of the Swift-Hohenberg equation, which was first used as a toy-model for the convective instability in the

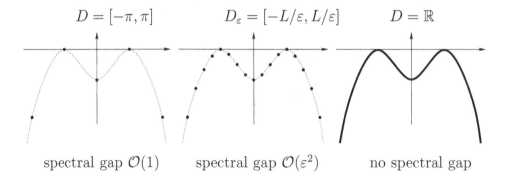

Fig. 1.1   Comparison between the spectrum of $-P(i\partial_x)$ for different domains. The dashed line is $k \mapsto -P(-k) = -(1-k^2)^2$, with dots, provided the corresponding e^{ikx} is an eigenfunction of $-P(i\partial_x)$.

Rayleigh-Bénard problem (see [SH77]). Here

$$u(t,x) \in \mathbb{R}, \quad \text{for} \quad t > 0, \ x \in D_\varepsilon = L\varepsilon^{-1} \cdot [-1,1]$$

fulfils

$$\partial_t u = -P(i\partial_x)u + \varepsilon^2 \nu u - u^3 + \varepsilon^{\frac{3}{2}}\xi \tag{1.14}$$

subject to periodic boundary conditions. Note that we prescribe a scaling between the noise strength and the distance from bifurcation, that differs from the one used in the bounded domain case.

The linear operator is given by

$$P(\zeta) = (1 - \zeta^2)^2 \ .$$

The complex eigenfunctions of the linear operator $P(i\partial_x)$ are $x \mapsto \exp\{ik\varepsilon\pi x/L\}$ with corresponding eigenvalue $P(k\varepsilon\pi/L)$ for $k \in \mathbb{Z}$. For simplicity, let ξ be space-time white noise in the following formal calculation. We rely on scaling properties for the noise, which are not that easy to formulate for coloured noise. See also Section 4.2. To be more precise, we use that ξ and $\hat{\xi}$ are versions of the same noise, when we define

$$\hat{\xi}(T,X) = \varepsilon^{-3/2}\xi(T\varepsilon^{-2}, X\varepsilon^{-1}) \ . \tag{1.15}$$

We expect a linear instability at $e^{\pm ix}$, as $P(\pm 1) = 0$ and $P(x) > 0$ for $x \neq \pm 1$, but due to the boundedness of the domain $e^{\pm ix}$ is in general not an eigenfunction. The nearest eigenfunction is $e^{i\rho_c(\varepsilon/L)x}$, where

$$\rho_c(\varepsilon/L) := \frac{\varepsilon\pi}{L} \cdot \left[\frac{L}{\varepsilon\pi}\right] ,$$

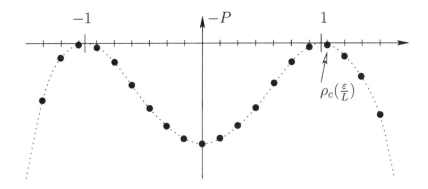

Fig. 1.2 Sketch of the eigenvalue curve for large domains of size $\mathcal{O}(\varepsilon^{-1})$. Note that $e^{\pm ix}$ is in general not an eigenfunction. The eigenfunctions with eigenvalue nearest to 0 are $e^{\pm i\rho_c x}$.

and $\left[\frac{L}{\varepsilon\pi}\right]$ is the nearest integer to $L/\varepsilon\pi$. See Figure 1.2, where we give a sketch of the eigenvalue curve of $-P(i\partial_x)$. Note that obviously

$$\left|\frac{1-\rho_c(\varepsilon/L)}{\varepsilon}\right| \le \frac{\pi}{2}\cdot\frac{1}{L}\,.$$

Leaving out the error term for simplicity of presentation, we make the following ansatz:

$$u(t,x) = \varepsilon A(\varepsilon^2 t, \varepsilon x)e^{i\rho_c(\varepsilon/L)x} + \varepsilon^3 B(\varepsilon^2 t, \varepsilon x)e^{3i\rho_c(\varepsilon/L)x} + c.c.\,, \qquad (1.16)$$

where $c.c.$ denotes the complex conjugate. Here $\Re(z) = \frac{1}{2}(z + c.c.)$ denotes the real part of $z \in \mathbb{C}$. The term involving B just simplifies the formal calculation. It has no impact on the final result. A similar idea was used in [KSM92] for the deterministic equation on the unbounded domain.

For the deterministic equation the effect of bounded but large domains was first discussed in [MSZ00]. They used the ansatz with $\rho_c \equiv 1$, and for the Swift-Hohenberg equation they obtain the following amplitude equation for A:

$$\partial_T A = 4\partial_X^2 A + \nu A - 3|A|^2 A\,, \qquad (1.17)$$

but subject to ε-dependent boundary conditions on $[-L, L]$, which take into account that the ansatz as a solution should be L/ε-periodic, which $e^{\pm ix}$ is in general not. We use the other ansatz (1.16) to see that all ε-dependent terms in the amplitude equation are actually uniformly small in L, and they vanish for $L \in \varepsilon\pi\mathbb{N}$. Moreover, due to (1.16) the formal calculation yields for the amplitude equation standard periodic boundary conditions, which are more familiar.

Plugging the ansatz (1.16) into (1.14) and using

$$-P(i\partial_x)[fe^{ikx}] = [-P(i\partial_x - k)f]e^{ikx}\,,$$

which is easy to verify, we get in lowest order (which is ε^3)

$$\partial_T A e^{i\rho_c x} + c.c. = \left[4\partial_X^2 A - 4i\frac{1-\rho_c^2}{\varepsilon}\partial_X A - \frac{(1-\rho_c^2)^2}{\varepsilon^2}A + \nu A \right] \cdot e^{i\rho_c x}$$
$$- (1-9\rho_c^2)^2 B e^{3i\rho_c x} - A^3 e^{3i\rho_c x} - 3A|A|^2 e^{i\rho_c x} + c.c. + \hat{\xi} .$$

Here $\hat{\xi}$, as in (1.15), is a rescaled version of ξ. In order to get rid of the terms depending on e^{3ix}, we choose

$$B = -(1-9\rho_c^2)^{-2} A^3 .$$

Furthermore, we use

$$1 - \rho_c^2 = 2(1-\rho_c) + \mathcal{O}\left(\frac{\varepsilon^2}{L^2}\right) .$$

Finally, we have to rewrite the noise. We will see below, that one can define a complex-valued space-time white noise η with law independent of ε, such that we obtain the following amplitude equation

$$\partial_T A = 4\left(\partial_X A - 2i\frac{1-\rho_c}{\varepsilon}\right)^2 A + \nu A - 3A|A|^2 + \eta , \qquad (1.18)$$

subject to periodic boundary conditions on $[-L, L]$. Note that $(1 - \rho_c)/\varepsilon$ is small for large L uniformly with respect to ε. Moreover, it vanishes for $L \in \varepsilon\pi\mathbb{N}$.

Discussion on the Noise

On the formal level there is no trivial argument available, why we can rewrite the noise ξ in order to obtain η. Furthermore, one has to be extremely careful with the formal analysis at this point. For example, consider the argument that there is a complex valued space-time white noise $\tilde{\eta}$ such that

$$\hat{\xi}(T, X) = \frac{1}{\sqrt{2}}\left[\tilde{\eta}(T, X)e^{ix} + c.c.\right] .$$

This is obviously true, by calculation the corresponding correlation functions. But the rigorous proofs of Chapter 4 show that this simple argument is wrong. Using $\tilde{\eta}$ we obtain a factor $1/\sqrt{2}$ in front of the noise in the amplitude equation, but this is not present in the rigorous result.

The main fault is that space-time white noise is of order $\mathcal{O}(\varepsilon)$, when we restrict it to modes in Fourier space of order larger than ε^{-1}. Especially, if we look at the linear equation, then it is possible to argue that modes are actually small, once their wavenumber is far away from the instability $\pm\rho_c$. Thus we neglect all modes that are sufficiently far away. Note that the instability $\pm\rho_c$ corresponds to modes e^{ikx} with wavenumber k of order $\pm\rho_c\pi/\varepsilon L$.

Thus after neglecting the modes far away from the instability, the right decomposition of ξ is now to separate it into two parts. One contains all the remaining

Fourier modes with positive wavenumber, and the other one all modes with negative wave-numbers. Then we can shift these modes in Fourier space by $\pm\rho_c\pi/\varepsilon L$ by pulling a factor $e^{\pm i\rho_c x}$ out of the Fourier series of the noise. Now we put these transformed noise processes into the amplitude equation, or its complex conjugate version.

Finally, we fill up these noise processes with noise on high wave numbers, in order to get space-time white noise again. The fact that we can fill up the high modes of the noise is by no means obvious from the PDE. The main reason is that we have a strong linear damping on high modes present in the Ginzburg-Landau equation (1.18), which arises from the linear part. See Section 4.3 or [BHP05]. It will be easier to see later, when we later change to the mild formulation.

1.2 General Structure of the Approach

It is not the aim of this section to present rigorous results. Instead it highlights the key steps in a non-technical way. For all our results in the stochastic case, the general method of proof already dates back to [BMPS01]. Furthermore it was already used for amplitude equations for deterministic equations, for instance, in [KSM92] and [Sch94].

For simplicity of presentation we focus on the case of bounded domains. The case of large or unbounded domains is similar, but it exhibits many additional technical difficulties. Furthermore, we stick to cubic nonlinearities with additive noise. This was discussed in Section 1.1.1. The method of proof for other types of equations is very similar, only the formulation and the technical details differ.

Due to the lack of regularity, we cannot proceed analogous to the deterministic setting. This is one of the main issues for SPDEs, as the approach for deterministic PDEs relies on bounds for solutions of the amplitude equations in spaces with sufficiently high regularity. But especially on large domains for SPDEs this is never the case. See Section 4.3 or Remark 4.1.

In order to give SPDEs like (1.2) a meaning, we use the concept of mild solutions. These are stochastic processes with continuous paths that fulfil the following variation of constants formula

$$u(t) = e^{tL}u(0) + \int_0^t e^{(t-\tau)L}[\varepsilon^2 Au + \mathcal{F}(u)](\tau)d\tau + \varepsilon^2 W_L(t) \qquad (1.19)$$

for $t \leq t^*$, where $t^* > 0$ is some stopping time. Here $\{e^{tL}\}_{t\geq 0}$ denotes the semigroup of operators generated by the differential operator L. For a detailed definition see [Paz83; Hen81; Lun95] or Section 2.5.1. The main point here is that $w(t) = e^{tL}w_0$ solves $\partial_t w = Lw$ with $w(0) = w_0$, and thus $\partial_t e^{tL} = Le^{tL}$.

For the definition of the stochastic convolution

$$W_L(t) = \int_0^t e^{(t-\tau)L}dQW(\tau) , \quad t \geq 0$$

see [DPZ92]. Formally differentiating (1.19) yields immediately that $u(t)$ solves (1.2).

Here $\partial_t QW = \xi$ in a generalised sense, and W is some cylindrical Wiener process in some Hilbert space (see Assumption 2.8 and the discussion below that). For the connection between the noise ξ and Q-Wiener processes see [Blö05b]. For a different approach using the Brownian sheet and an explicit representation of the semigroup e^{tL} via the Green function see [Wal86].

We use the projection P_c onto the kernel $\mathcal{N}$ of L and $P_s = I - P_c$, which were defined before (cf. Section 1.1.1). Now we project the equation to $\mathcal{N}$ and $\mathcal{S}$.

Definition 1.1   We call $u_s(t) = P_s u(t) \in \mathcal{S}$ *fast modes*, as they are subject to a deterministic exponential decay on a time-scale of order $\mathcal{O}(1)$. Moreover, $u_c(t) = P_c u(t) \in \mathcal{N}$ are the *slow modes*, as they change only on the slow time-scale $T = \varepsilon^2 t$.

For simplicity we assume that P_c, and hence P_s, commutes with L and therefore with e^{tL}, too. Moreover, $e^{tL} P_c = P_c = P_c e^{tL}$. Projecting (1.19), we derive

$$u_c(t) = u_c(0) + \int_0^t [\varepsilon^2 A_c(u_c + u_s) + \mathcal{F}_c(u_c + u_s)](\tau)d\tau + \varepsilon^2 P_c QW(t) , \qquad (1.20)$$

and

$$u_s(t) = e^{tL} u_s(0) + \int_0^t e^{(t-\tau)L}[\varepsilon^2 A_s(u_c + u_s) + \mathcal{F}_s(u_c + u_s)](\tau)d\tau + \varepsilon^2 P_s QW_L(t). \qquad (1.21)$$

If we now use the scaling

$$u_c = \mathcal{O}(\varepsilon) \quad \text{and} \quad u_s = \mathcal{O}(\varepsilon^2) ,$$

as indicated in the ansatz (1.4), then we immediately see that (1.20) decouples from (1.21). Here we just neglect all terms of higher order and rescale to the slow time-scale. Again we derive the amplitude equation (1.5), but now in integrated form.

The Impact of Noise

We see in (1.20) and (1.21) that the influence of the noise on the slow and fast component is very different. On the time-scale t of order 1 both are ε^2, but this changes significantly, when using the slow time-scale $T = \varepsilon^2 t$. In (1.20) for the slow variable u_c, we see that the noise acts like some Brownian motion in $\mathcal{N}$, which is isomorphic to $\mathbb{R}^{\dim(\mathcal{N})}$. Due to the scaling properties of W it is possible to verify that

$$\sup_{t \in [0, T_0 \varepsilon^{-2}]} \|P_c QW(t)\| = \varepsilon^{-1} \cdot \sup_{T \in [0, T_0]} \|P_c QW(T)\| \quad \text{(in law)} .$$

Hence,

$$\sup_{t\in[0,T_0\varepsilon^{-2}]} \|P_cQW(t)\| = \mathcal{O}(\varepsilon^{-1})\,.$$

On the other hand, in (1.21) for the fast component u_s the noise enters as an infinite dimensional Ornstein-Uhlenbeck process. Here it is possible to verify using the celebrated factorisation method (see [DPZ92] or the proof of Lemma A.4) that

$$\sup_{t\in[0,T_0\varepsilon^{-2}]} \|P_sW_L(t)\| = \mathcal{O}(\varepsilon^{-\kappa}) \quad \text{for arbitrary } \kappa > 0\,.$$

See for instance Theorem 5.1 of [BMPS01]. Note that this is not the optimal bound. The right scaling should be logarithmic in ε.

From the previous discussion we see that the noise acting on (1.21) is about one order of magnitude smaller than the noise acting on (1.20). Using the stability of the equations a long calculation leads to a priori estimates. These establish different scalings for u_c and u_s, which are the ones already used in the ansatz (1.4). This is the so called *attractivity result* (see Theorem 1.1), which justifies the ansatz for the formal calculation. Moreover, it provides the typical structure of the initial condition necessary to start an *approximation result*. This means a rigorous error bound between the true solution, and the approximation via amplitude equations. See Theorem 1.3.

1.2.1 *Meta Theorems*

The first result presented here is the attractivity. It justifies the scaling of ansatz (1.4) used for the formal derivation. It heavily relies on the structure of the equation. Sometimes we rely on global nonlinear stability and sometimes we only use linear stability on the non-dominant modes. A typical statement would be:

Theorem 1.1 **(Attractivity)** *There is a time $t_\varepsilon = \mathcal{O}(\ln(\varepsilon^{-1}))$ such that for all solutions u of (1.19) with initial conditions $u(0)$ of order $\mathcal{O}(\varepsilon)$ we have $u_s(t_\varepsilon) = \mathcal{O}(\varepsilon^2)$ and $u_c(t_\varepsilon) = \mathcal{O}(\varepsilon)$. This means the solution looks at the time t_ε like ansatz (1.4). To be more precise $u(t_\varepsilon) = \varepsilon a_\varepsilon + \varepsilon^2\psi_\varepsilon$ with $a_\varepsilon \in \mathcal{N}$ and $\psi_\varepsilon \in \mathcal{S}$ both of order $\mathcal{O}(1)$.*

If we assume additionally global nonlinear stability for the equation, then there is a time $T_\varepsilon = \mathcal{O}(\varepsilon^{-2})$ such that $u(T_\varepsilon) = \mathcal{O}(\varepsilon)$ independent of the initial condition.

This theorem is rigorously stated in Theorems 2.7 or 2.8. We will give a detailed discussion of these results for multiplicative noise in Theorems 2.1 and 2.4 for cubic and quadratic nonlinearities. A sketch of the typical dynamics for the local attractivity result is given in Figure 1.3.

Remark 1.4 *Depending on the assumptions the statement $g_\varepsilon = \mathcal{O}(f_\varepsilon)$ can have two different meanings. Depending on the context, we either use that for all $p > 0$ there is a constant $C > 0$ such that $\mathbb{E}\|g_\varepsilon\|^p \leq Cf_\varepsilon^p$ for all $\varepsilon \in (0,1]$. This is typically*

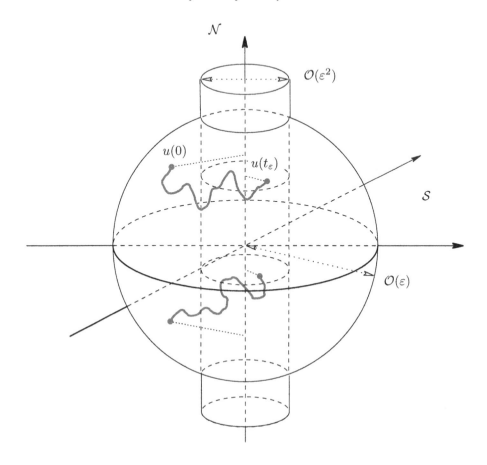

Fig. 1.3   Two typical trajectories for the local attractivity result from Theorem 1.1. For the global result relying on nonlinear stability the $\mathcal{O}(\varepsilon)$-ball would be attractive, too.

only valid for nonlinear stable equations, where we can actually bound moments. In case of, for instance quadratic nonlinearities, where in general we do not have control on moments of solutions, we also use the somewhat weaker meaning that for some constant $C > 0$, we have $\mathbb{P}(\|g_\varepsilon\| \geq Cf_\varepsilon)$ uniformly small for all $\varepsilon \in (0, 1]$. Sometimes we also give explicit convergence rates of this probability for $\varepsilon \to 0$.

For a solution a of (1.5) and ψ of (1.6) we define first the approximations εw_k of order k by

$$\varepsilon w_1(t) := \varepsilon a(\varepsilon^2 t) \quad \text{and} \quad \varepsilon w_2(t) := \varepsilon a(\varepsilon^2 t) + \varepsilon^2 \psi(t) . \tag{1.22}$$

In our setting the *residual* of εw is defined by

$$\text{Res}(\varepsilon w)(t) = -\varepsilon w(t) + e^{tL}\varepsilon w(0) + \varepsilon^2 W_L(t) \tag{1.23}$$

$$+ \int_0^t e^{(t-\tau)L}[\varepsilon^3 Aw + \mathcal{F}(\varepsilon w)](\tau)d\tau.$$

In order to show that εw_j is a good approximation of a solution u of (1.19), the key step is to control the residual, which measures the distance of the approximation εw_j from being a solution. Obviously, $\mathrm{Res}(\varepsilon w_j) = 0$, if and only if εw_j is a solution of (1.19).

Note that in general no additional information on the solution u is needed. We mainly rely on a priori estimates for the solutions of the amplitude equation, which are much easier to obtain. A typical statement is:

Theorem 1.2 (Residual) *For any $T_0 > 0$, and approximations given by (1.22) with initial conditions $a(0)$ and $\psi(0)$ of order $\mathcal{O}(1)$, we have*

$$\sup_{t \in [0, T_0 \varepsilon^{-2}]} \|\mathrm{Res}(\varepsilon w_k)(t)\| = \mathcal{O}(\varepsilon^{1+k}).$$

A detailed discussion of the residual in the multiplicative noise case can be found in Theorem 2.2 for cubic and in Theorem 2.5 for quadratic nonlinearities.

Using the results for the residual, it is usually straightforward to derive the approximation result. Here we sometimes have to use additional assumptions on the nonlinearity. Again no bounds on solutions are necessary, but for simplicity we sometimes rely on them, when they are easy to establish.

Theorem 1.3 (Approximation) *For all $T_0 > 0$, all solutions u of (1.19), and all approximations given by (1.22) with $\|u(0) - \varepsilon w_k(0)\| = \mathcal{O}(\varepsilon^{1+k})$ we have*

$$\sup_{t \in [0, T_0 \varepsilon^{-2}]} \|u(t) - \varepsilon w_k(t)\| = \mathcal{O}(\varepsilon^{1+k}).$$

This result is rigorously stated in Theorems 2.9 and 2.10. A detailed discussion for multiplicative noise is given in Theorems 2.3 or 2.6. We sketch a typical trajectory from the approximation result in Figure 1.4. Note finally that the condition imposed on the initial conditions in Theorem 1.3 are usually trivial, in case we consider a solution given by the attractivity result.

1.3 Examples of Equations

In the literature there are numerous examples of equations where the abstract theorems do apply. In this section we focus mainly on additive noise. For instance, for cubic nonlinearities the well known Ginzburg-Landau equation (see [DE00] for a standard proof of existence of unique solutions)

$$\partial_t u = \Delta u + \nu u - u^3 + \sigma \xi$$

and the Swift-Hohenberg equation (see [CH93] for numerous references)

$$\partial_t u = -(\Delta + 1)^2 u + \nu u - u^3 + \sigma \xi$$

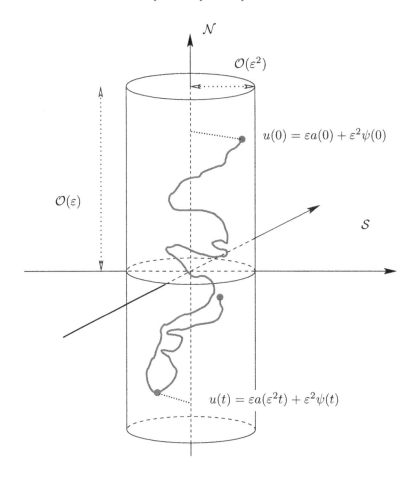

Fig. 1.4   A typical trajectory for the approximation result.

fall into the scope of our work, in case the parameters ν and σ are small and of comparable order of magnitude. Both equations are considered on bounded domains with suitable boundary conditions (e.g. periodic, Dirichlet, Neumann, etc.). The Swift-Hohenberg equation is a toy model for the convective instability in the Rayleigh-Bénard convection. A formal derivation of the equation from the Boussinesq approximation of fluid dynamics can be found in [SH77].

Another example arising in the theory of surface growth is

$$\partial_t u = -\Delta^2 u - \mu \Delta u + \nabla \cdot \left(|\nabla u|^2 \nabla u \right) + \sigma \xi \,, \tag{1.24}$$

subject to periodic boundary conditions and moving frame $\int_G u \, dx = 0$, where one rescales the mean growth of u out of the equation, in order to ensure a Poincare type inequality. This model was first suggested by [LDS91]. The deterministic equation was rigorously treated in [KSW03]. For a review on surface growth see for example [BS95] or [HHZ95]. For this model we can consider $\mu = \mu_0 + \varepsilon^2$ and $\sigma = \mathcal{O}(\varepsilon^2)$, where μ_0 is such that $L = -\Delta^2 u - \mu_0 \Delta u$ is a non-positive operator with non-zero kernel. We will see later on, that all examples presented up to now exhibit a stable nonlinearity in the sense of Assumption 2.2.

There are also numerous examples in the physics literature of equations with quadratic nonlinearities where our theory does apply. One example is the growth of rough amorphous surfaces. See for example [BGR02], [BG02b], or [BH04] and the references therein. The equation is of the type

$$\partial_t u = -\Delta^2 u - \mu \Delta u - \Delta |\nabla u|^2 + \sigma \xi , \qquad (1.25)$$

for instance subject to periodic boundary conditions on $[0, L]^d$ and the moving frame condition. In the context of surface growth, this equation was first used in [SP94]. Later on it was used to model a special class of growth of amorphous surfaces (See for example [RML$^+$00; RLH00] and the references therein). The local existence of unique mild solutions using standard fixed point methods is verified in [BG04] or for the deterministic equation in [SW05]. The uniqueness of global solutions is still open, even for $d = 1$. For the existence of global solutions with Markov property see [BFR06].

Other examples that fall into the scope of our work are first the Kuramoto-Sivashinsky equation

$$\partial_t u = -\Delta^2 u + \mu \Delta u + |\nabla u|^2 + \sigma \xi .$$

This was first used to model propagation of flames and recently as a model for surface growth via ion sputtering (cf. [CB95; FBK$^+$02; LV05]). The second example is the Burgers' equation which was already discussed in Section 1.1.3. Another related model is the celebrated KPZ-equation.

But the probably most important example is the Rayleigh-Bénard problem (see for example [Get98; Wal97] or [CH93]) which is the paradigm of pattern formation in convection problems. In [Blö05a] we discuss in detail the amplitude equation for the stochastic two-dimensional Bénard problem in a strip, where a fluid is heated from below. The three dimensional problem in a box is treated similarly, but the notation is much more involved.

In the following denote by (v, w) the velocity field of the fluid in $(y, z) \in D :=$ $[0, 2\pi] \times [0, \pi]$, where z is the vertical direction. Hence, the fluid is heated at $z \equiv 0$. Let p be the pressure and θ the normalised temperature, which means that $\theta \equiv 0$ and $(v, w) \equiv 0$ is heat transport without motion.

In dimensionless quantities the governing Navier-Stokes and heat equation are given by (see e.g. [Get98] or [Wal97])

$$\partial_t(v,w) + ((v,w) \cdot \nabla)(v,w)) = -\nabla p + (0,1)\frac{R}{P}\theta + \Delta(v,w) \qquad (1.26)$$

$$\partial_t\theta - v + ((v,w) \cdot \nabla)\theta = \frac{1}{P}\Delta\theta + \varepsilon^2\xi \qquad (1.27)$$

$$\mathrm{div}(v,w) = 0 . \qquad (1.28)$$

We suppose periodic boundary conditions in y both for θ and (v,w). Moreover, $\partial_z v = w = \theta = 0$ for $z = 0$ and $z = \pi$. The noise ξ corresponds to fluctuations in the temperature. We could also incorporate fluctuations in the velocity field, but we neglect this for simplicity of presentation.

In order to rule out motion of the whole fluid in the y-direction, we suppose vanishing mean flux $\int_0^\pi v dz$. We use the following constants:

- R the Rayleigh number
- P the Prantl number
- $\rho = R/P$ the Reynolds number

The Rayleigh number is a dimensionless measure of the heat difference between top and bottom of the strip, while the Prantl number depends only on the properties of the fluid.

For a rigorous verification of the amplitude equation for the Bénard problem see [Blö05a]. We give a summary of all these results in Section 2.6. Note that in this case one of the major difficulties is that the linear operator L is not self-adjoint. However it exhibits a complete non-orthogonal basis of eigenfunctions. For simplicity of presentation, we do not focus on that technical point (cf. Assumption 2.1). For a detailed discussion see [Blö05a].

In this example L and A do not commute, and hence $P_c A \neq A P_c$, which is in contrast to most of the other examples stated above, where we have equality. This does not cause major technical difficulties, and we allow for quite general A in the abstract setting (cf. Assumptions 2.2 or 2.4).

Chapter 2

Amplitude Equations on Bounded Domains

On bounded domains, we can approximate on long time-scales the essential dynamics of an SPDE near a change of stability by the amplitude equation. This is in this chapter just an SDE describing the dynamics of the dominating modes, which are the ones that change sign in the linearisation. For the formal derivation in the case of additive noise see Sections 1.1.1 or 1.1.3. The main mathematical reason why the other modes are not important is the presence of a well defined spectral gap in the linearised equation of order $\mathcal{O}(1)$ between the eigenvalues of the dominant eigenfunctions and the remainder of the spectrum.

The approximation via SDE is only meaningful for small domains. If the domain gets larger, one needs very small noise to apply the results. See Chapter 4, where the size of the domain is coupled to the distance from bifurcation. Problems arise due to the fact that, if we enlarge the domain, then we shrink the spectral gap. The precise definition of the spectral gap ω will be given in Assumption 2.1. The main problem is that a lot of constants depend on ω, and they tend to infinity for $\omega \to 0$. But if the domain-size $\ell \to \infty$, then in most cases $\omega \to 0$. Hence, for large domains our result is only meaningful for very small noise strength ε^2 with $\varepsilon \in (0, \varepsilon_0]$, where $\varepsilon_0 = \varepsilon_0(\ell) \to 0$ for $\ell \to \infty$. However, the linear part of our equation is usually coupled to the noise, and thus has to be very small, too. The main problem is now, that this linear part reflects the influence of control parameters adjusted in experiments. It is not possible to consider it arbitrarily small.

In the following, we demonstrate the power of our approach by applying it to PDEs perturbed by a simple multiplicative noise. Although our results apply to more complicated noise terms, for simplicity of presentation we consider only this very simple example in order to outline the main ideas in a less technical way. The results for additive noise are reviewed later in this chapter.

Multiplicative noise arises, for instance, in models with noisy bifurcation parameters. For more details about multiplicative noise see Section 2.1. The proofs will be somewhat easier, as Itô's formula and Burkholder-Davis-Gundy type inequalities (cf. Section A.1) are available, which one cannot use easily for additive space-time white noise. Nevertheless, we have to modify some results, for example, the Burkholder-Davis-Gundy type inequalities. We need precise dependence of the constants on various parameters, which was not derived before. This is presented in the Appendix A.

This section is organised as follows. Sections 2.2 and 2.4 state and prove our main results for multiplicative noise, while Section 2.3 provides technical, but straightforward a priori estimates. In Section 2.5 we review the results on additive noise, which were derived for cubic nonlinearities first in [BMPS01] and they were later extended in [Blö03; BH04; BH05]. For quadratic nonlinearities, which also covers the Rayleigh-Bénard convection, we present a summary of the results of [Blö05a] in Section 2.6.

2.1 Multiplicative Noise (Parameter Noise)

Let us first motivate, why we are interested in multiplicative noise. It appears naturally in models, where one considers noisy control parameters. Consider as an example some deterministic PDE of the following type

$$\partial_t u = Lu + \mu u + \mathcal{F}(u) , \qquad (2.1)$$

where L is some linear differential operator and $\mathcal{F}$ is some nonlinearity, for instance $-u^3$. Suppose that the equation undergoes a change of stability (or bifurcation) when $\mu = 0$.

The question is, whether we can see the influence of small noise in the bifurcation parameter μ in the case where μ is near or at the bifurcation. This is an important question in many experiments, as μ models experimental quantities like, for instance, temperature, which are naturally subject to small (random) perturbations.

We consider in (2.1) a simplified PDE model, where the perturbation of the parameter has no spatial dependence and is homogeneous in space. This kind of equation was recently studied in more detail, for instance, by [CLR00; CLR01; Rob02] where they determined the dimension and structure of a random attractor for a stochastic Ginzburg-Landau equation. On the other hand, even the stability of linear equations (i.e. $\mathcal{F} \equiv 0$) was only studied recently in [CR04] or [Kwi02] following the celebrated work of [ACW83].

Let us come back to (2.1). Assume that the control parameter $\mu \in \mathbb{R}$ is perturbed by white noise and suppose the strength of the fluctuations $\varepsilon > 0$ is small. A typical model is a Gaussian noise μ with some mean and covariance functional

$$\mathbb{E}\mu(t) = \mu_\varepsilon \in \mathbb{R}, \qquad \mathbb{E}(\mu(t) - \mu_\varepsilon)(\mu(s) - \mu_\varepsilon) = \varepsilon^2 \delta(t - s) .$$

Thus we can write $\mu = \mu_\varepsilon + \varepsilon\xi$, where $\xi = \partial_t \beta$ is the generalised derivative of a real valued Brownian motion $\beta = \{\beta(t)\}_{t \geq 0}$.

Hence, we can rewrite (2.1) as a stochastic PDE

$$\partial_t u = Lu + \mu_\varepsilon u + \mathcal{F}(u) + \varepsilon u \partial_t \beta . \qquad (2.2)$$

Note that for our considerations it is necessary to be sufficiently close to the bifurcation, in order to see both the influence of the linear instability μ_ε and the noise $\varepsilon\xi$ in the amplitude equation. Using scaling arguments we will see later on that $\mu_\varepsilon = \mathcal{O}(\varepsilon^2)$ is necessary for both effects appearing. Larger μ_ε leads to the loss of the noise in the amplitude equation, while smaller μ_ε leads to the loss of the linear term.

Stratonovič vs. Itô

There is a well known problem in the interpretation of equation (2.2). It has different meanings, if we consider the term $u\partial_t\beta$ for instance in the Itô or Stratonovič sense. In order to have Itô's formula at hand, we will focus on the Itô interpretation. But we can easily use the Stratonovič interpretation, as the Stratonovič-to-Itô correction only gives terms that are linear in u. This is well known. See for example [Øks98] for an SDE setting. The linear term from the correction appearing in (2.2) is u times a constant of order $\mathcal{O}(\varepsilon^2)$. We can treat this as a modification of the term $\mu_\varepsilon u$. Then we can apply the approximation, and use the Itô-to-Stratonovič correction for the amplitude equation. In the end, we would get the same result with all Itô differentials replaced by Stratonovič differentials.

Other Multiplicative Noise

In the following we look for simplicity of presentation only at our toy model given by (2.2). It is straightforward to generalise the approach presented for the simple one-dimensional noise in (2.3) to noise terms like $\varepsilon \sum_{k=1}^N \mathcal{L}^{(k)} u d\beta_k$, where the $\{\mathcal{L}^{(k)}\}_{k=1,\dots,N}$ are linear operators, and the $\{\beta_k\}_{k=1,\dots,N}$ are independent Brownian motions. Each of the terms in the sum now gives a contribution of the type $P_c \mathcal{L}^{(k)} a d\tilde{\beta}_k$ to the amplitude equation. For simplicity of presentation, we work with the simpler model.

More realistic noise depending on space and time was treated for instance in [Cer05], where the long-time behaviour in terms of invariant measures was studied. The work relies on the use of comparison principles, which are mainly available for L being a second order elliptic differential operator.

For the precise mathematical model of multiplicative space-time white noise see [DPZ92] or [Wal86]. There one defines multiplicative noise $u\partial_t W$, where $\partial_t W$ is space-time white noise given by the generalised derivative of some standard cylindrical Wiener-process

$$W(t) = \sum_{k\in\mathbb{N}} \beta_k(t)e_k \ ,$$

with

- $\{\beta_k\}_{k\in\mathbb{N}}$ independent identically distributed standard Brownian motions
- $\{e_k\}_{k\in\mathbb{N}}$ any L^2-orthonormal basis

A major problem is to evaluate the projection of the multiplicative noise term onto the kernel $\mathcal{N}$ of L. The main problem being the terms $P_c(ue_k)$, as this depends heavily on the chosen basis. Using the ansatz

$$u(t) = \varepsilon a(\varepsilon^2 t) + \mathcal{O}(\varepsilon^2) \,,$$

it is possible to compute the noise term in the amplitude equation for $a \in \mathcal{N}$. Formally, we should get a term

$$P_c[a\partial_T \hat{W}(T)] = \sum_{k \in \mathbb{N}} P_c(ae_k)\partial_T \hat{\beta}_k(T) \,,$$

where $\hat{\beta}_k(T) = \varepsilon\beta_k(\varepsilon^{-2}T)$ are rescaled Brownian motions. Depending on the orthonormal basis, this term could be hard to evaluate. It is possible that these sums contain only finitely many terms, but in general we need summability conditions.

We could also treat more complicated nonlinearities. Using cut-off techniques and Taylor expansions of the nonlinearities, we can always reduce these equations to simpler models treated here. Keeping in mind that we are looking only for small solutions in a small neighbourhood of the change of stability. Nevertheless the results may change if in lowest order the nonlinearity is not cubic or quadratic, or the parameter is coupled to a quadratic function in u.

The results presented in this section are based on my Habilitation and partly on [BHP07]. However, the general method is of course similar to older results like [Blö03; Blö05a; BH04] or [BMPS01]. See Section 1.2 for an outline of the general strategy of proof. Some improvement in technical results are also presented, that will also improve these older results. The main differences are on one hand, that we simplify some parts of the proofs using Itô's formula, but on the other hand, we are no longer able to just use deterministic results path-wise as in [BMPS01; Blö05a] or [Blö03]. One needs more probabilistic tools in our proofs. Finally we would like to point out again, that for simplicity of presentation we do not aim for the highest generality possible, as even this simple setting the results are already quite technical.

2.2 Assumptions and Results — The Cubic Case

This section summarises for the cubic case all assumptions necessary and states the main results. In this chapter we treat two sets of different assumptions. On one hand this section treats nonlinear stable equations involving cubic terms, where we can use standard a priori estimates to obtain bounds on moments of solutions. On the other hand we consider in Section 2.4 quadratic nonlinearities, which in general do not allow to bound moments of solutions. Especially, if we cannot rule out the possibility of a blow-up of solutions in finite time, which is the case in many examples. One is the 2D Kuramoto-Sivashinsky equation, for instance. In this case we obtain local result by using cut-off techniques.

Consider the following SPDE in some Hilbert space X with scalar product $\langle \cdot, \cdot \rangle$ and norm $\| \cdot \|$. We could also consider Banach spaces here, but the Hilbert space setting simplifies the notation and the a priori estimates on solutions.

$$du = [Lu + \varepsilon^2 Au + \mathcal{F}(u)]dt + \varepsilon u d\beta . \tag{2.3}$$

The precise setting is given below in Assumptions 2.1 for L, 2.2 for A and $\mathcal{F}$, and 2.3 for the Itô-differential.

Assumption 2.1 *Suppose L is the infinitesimal generator of some analytic semigroup $\{e^{tL}\}_{t \geq 0}$ in X. Suppose the kernel $\mathcal{N} := N(L)$ is finite dimensional, and the projection P_c onto $\mathcal{N}$ commutes with L. Define $P_s = I - P_c$.*
Suppose that there are constants $M > 0$ and $\omega > 0$ such that for all $t > 0$

$$\|e^{tL}\|_{\mathcal{L}(X,X)} \leq M \quad \text{and} \quad \|P_s e^{tL}\|_{\mathcal{L}(X,X)} \leq M e^{-t\omega} . \tag{2.4}$$

Definition 2.1 Denote by $\mathcal{L}(X,Y)$ the space of continuous linear operators between the spaces X and Y with norm

$$\|T\|_{\mathcal{L}(X,Y)} = \sup\{\|Tu\|_Y : \|u\|_X = 1\} .$$

Furthermore, $\mathcal{L}_k(X,Y)$ denotes the space of continuous k-linear mappings from $X \times \cdots \times X$ to Y.

As the dimension of $\mathcal{N}$ is finite, it is well known that both P_c and P_s are bounded linear operators on X (cf. [Wei80]). This means that P_c, $P_s \in \mathcal{L}(X) = \mathcal{L}(X,X)$. Note that in general it is not necessary that P_c is self-adjoint. Especially if L is not self-adjoint, then the projection need not be orthogonal.

Definition 2.2 Let the operator L be given by Assumption 2.1 Define the fractional space X^α for $\alpha \geq 0$ as the domain of definition

$$X^\alpha = D((1-L)^\alpha) \quad \text{with norm} \quad \|u\|_\alpha = \|(1-L)^\alpha u\| .$$

The space $X^{-\alpha} = (X^\alpha)'$ is the dual of X^α with respect to the pairing $\langle \cdot, \cdot \rangle$ induced by the scalar product in X.

For details see for example [Lun95; Hen81] or [Paz83]. It is well known that $\{e^{tL}\}_{t \geq 0}$ extends to an analytic semigroup on all X^α, $\alpha \in \mathbb{R}$. Note furthermore that obviously $\mathcal{N} \subset X^\alpha$ for all $\alpha \geq 0$, since $(1-L)^\alpha \mathcal{N} = \mathcal{N}$.

Under Assumption 2.1 it is also a well known fact that for all $\alpha \geq 0$ there is a constant $M_\alpha \geq 1$ such that

$$\|e^{tL}\|_{\mathcal{L}(X^{-\alpha},X)} \leq M_\alpha(1 + t^{-\alpha}) \quad \text{for all} \quad t > 0 \tag{2.5}$$

and for some $0 < \tilde{\omega} < \omega$ and possibly different constants M_α

$$\|P_s e^{tL}\|_{\mathcal{L}(X^{-\alpha},X)} \leq M_\alpha(1 + t^{-\alpha})e^{-t\tilde{\omega}} \quad \text{for all} \quad t > 0 . \tag{2.6}$$

For the nonlinearity $\mathcal{F}$ in (2.3) we assume that it is of stable cubic type. The precise assumption on A and $\mathcal{F}$ is the following.

Assumption 2.2   *Let $A, \mathcal{F} : D(L) \to X$ such that $A \in \mathcal{L}(X, X^{-\alpha})$ for some $\alpha \in [0, 1)$ and that we can extend $\mathcal{F}$ to a continuous trilinear operator $\mathcal{F} \in \mathcal{L}_3(X, X^{-\alpha})$. Define for short-hand notation*

$$\mathcal{F}_c = P_c\mathcal{F}, \quad A_c = P_cA, \quad \text{and} \quad \mathcal{F}(u) = \mathcal{F}(u, u, u) .$$

We assume that the following strong nonlinear stability condition holds: There are constants $C, c > 0$ such that

$$\langle u, \varepsilon^2 Au + \mathcal{F}(u) + Lu \rangle \le C\varepsilon^4 - c\|u\|^4 \qquad \text{for all } u \in D(L) . \tag{2.7}$$

Furthermore let the following dissipativity condition of $\mathcal{F}$ in $\mathcal{N}$ be true:

$$\langle \mathcal{F}_c(a) - \mathcal{F}_c(b), a - b \rangle \le 0 \quad \text{for all } a, b \in \mathcal{N} . \tag{2.8}$$

Although we state it in the assumption, it is not necessary that $\mathcal{F}$ is a cubic operator. The main ingredient is the global dissipativity condition, and some conditions on locally cubic behaviour. However, we could consider higher order nonlinearities to obtain nonlinear stability. But we cannot easily get rid of (2.8), which more generally states that the cubic part of the nonlinearity projected onto $\mathcal{N}$ is dissipative. If the amplitude equation is nonlinearly unstable, then we encounter plenty of difficulties like blow up, for example.

Remark 2.1   *The Hilbert space setting is just for simplicity. Here we easily derive a priori estimates for the solutions. Instead we could use for instance conditions for the differential of the norm in some Banach space. We could also use a much more general setting, as for instance in [Blö03], where non-autonomous locally cubic nonlinearities were treated. Here we focus on the simpler setting in order to explain the main ideas.*

For the noise we suppose:

Assumption 2.3   *Let $d\beta$ be the Itô differential with respect to the real-valued standard Brownian motion $\{\beta(t)\}_{t \ge 0}$ adapted to some filtration $\{\mathcal{F}_t\}_{t \ge 0}$ on a probability space $(\Omega, \mathcal{A}, \mathbb{P})$.*

Again, we do not focus on the most general setting. Instead we use this simple type of noise, in order to outline the main ideas. In Section 2.1 we already outlined some of the problems arising for more realistic types of noise.

With this simple kind of noise it is straightforward to define stochastic integration of Hilbert-space valued functions with respect to $d\beta$. This is completely analogous to stochastic ODEs in $\mathbb{R}^n$ (cf. [Mao97; Øks98]).

To give a meaning to the SPDE (2.3) we consider mild solutions given by the following definition.

Definition 2.3 Let Assumptions 2.1, 2.2, and 2.3 be true. We call an X-valued stochastic process $\{u(t)\}_{t\geq 0}$ a mild solution of (2.3) in X, if it is adapted to the filtration $\{\mathcal{F}_t\}_{t\geq 0}$ and there is a positive stopping time $\tau_e > 0$ such that $u \in C^0([0, \tau_e), X)$ and the following variation of constants formula

$$u(t) = e^{tL}u(0) + \int_0^t e^{(t-\tau)L}[\varepsilon^2 Au + \mathcal{F}(u)](\tau)d\tau + \varepsilon \int_0^t e^{(t-\tau)L}u(\tau)d\beta(\tau) \quad (2.9)$$

holds for all $0 < t < \tau_e$. We choose $[0, \tau_e)$ as a maximal interval of existence. This means either

$$\tau_e = \infty \quad \text{or} \quad \|u(t)\| \to \infty \text{ for } t \to \tau_e .$$

Remark 2.2 *For simplicity of presentation we do not focus on the construction of continuous versions of u and equations fulfilled only $\mathbb{P}$-almost sure. We always suppose that we already have a version of u that is continuous and solves (2.9) for all realizations.*

Remark 2.3 *The assumption that u is adapted to the filtration $\{\mathcal{F}_t\}_{t\geq 0}$ is not important in the sequel. It follows immediately from (2.9), as $u(t)$ depends only on $\{\beta(s)\}_{s\in[0,t]}$.*

We also need the notion of a strong solution. To avoid technical problems, we suppose for simplicity that all strong solutions are mild solutions. This is not obvious from the definition, but it is easy to verify that a mild solution is strong, if we have enough regularity of solutions. This is just a technical issue.

Definition 2.4 Let Assumptions 2.1, 2.2, and 2.3 be true. We call a mild solution of (2.3) in X a strong solution in X, if

$$\mathbb{E}\int_0^t \|[Lu + \varepsilon^2 Au + \mathcal{F}(u)](\tau)\|d\tau < \infty, \qquad \mathbb{E}\int_0^t \|u(\tau)\|^2 d\tau < \infty \quad (2.10)$$

for all $t < \tau_e$ and

$$u(t) = u(0) + \int_0^t [Lu + \varepsilon^2 Au + \mathcal{F}(u)](\tau)d\tau + \varepsilon \int_0^t u(\tau)d\beta(\tau) \quad (2.11)$$

in X for all $t \in [0, \tau_e)$. Again we choose τ_e to be maximal. This means that either $\tau_e = \infty$ or one condition in (2.10) fails to be true at $t = \tau_e$.

This is slightly stronger definition than the one in [DPZ92], as we actually impose conditions on the moments to exist. However, this is mainly for simplicity of presentation. One can significantly weaken that and suppose only that instead of expectations only the probability of the terms in (2.10) being finite is 1.

We remark that the second condition in (2.10) is not necessary for our proofs. It is mainly for convenience, in order to have the stochastic integral in (2.11) well defined. Again, $\mathbb{P}(\int_0^t \|u(\tau)\|^2 d\tau < \infty) = 1$ would be sufficient.

Using standard theory given for instance in [DPZ92], it is easy to verify that under suitable assumptions (like Lipschitz continuity of $\mathcal{F}$, for instance) there is a unique mild solution in X with $\tau_e = \infty$. This relies on fixed point arguments.

Furthermore, using cut-off techniques, like for SDEs in Section 3.3 of [McK69] or Section 1.2 of [Cer01], it is easy to see that under our assumptions, there is a unique mild solution in X. The existence of strong solutions is standard under additional regularity conditions on $\mathcal{F}$, but we do not focus on that in detail.

For instance, under our assumptions, if we additionally assume that $A \in \mathcal{L}(X^1, X^{\tilde{\alpha}})$ and $\mathcal{F} \in \mathcal{L}_3(X^1, X^{\tilde{\alpha}})$ for some $\tilde{\alpha} > 0$, then it is possible to check, that given an initial value in X^1 there is a unique mild solution in X^1, which is also a strong solution in X. Nevertheless, the last step is not that straightforward, as we need a priori estimates on moments of the X^1-norm of the solution u. For simplicity of presentation later in Section 2.3 we focus on a priori bounds in the space X only. The bounds in X^1 are just technical generalisations of these results.

For the method of proof in the cubic case, we follow mainly the outline in Section 1.2. Although some parts of the proof are simplified, we encounter technical difficulties not present for additive noise. See for example [BMPS01] or [BH04].

2.2.1   *Attractivity*

We establish two results. The first one in Theorem 2.1 is a very strong result. It relies on the nonlinear stability of the equation and establishes bounds on $\mathbb{E}\|u(t)\|^p$ for large t completely independent of the initial condition $u(0)$. The second result is somewhat weaker. It relies on the existence of bounds on $\mathbb{E}\|u(0)\|^p$, and it establishes bounds on $\mathbb{E}\|P_s u(t)\|^p$ for moderately large t. This relies mainly on the linearised picture and a spectral gap of the linearised operator.

For the attractivity our main goal is to verify that there is a time $t_\varepsilon > 0$ such that

$$u(t_\varepsilon) = \varepsilon a_\varepsilon + \varepsilon^3 \psi_\varepsilon \,,$$

where $a_\varepsilon \in \mathcal{N}$ and $\psi_\varepsilon \in P_s X$ are both of order $\mathcal{O}(1)$.

Theorem 2.1   (Attractivity) *Let Assumptions 2.1, 2.2, and 2.3 be true and let u be a strong solution of (2.3) in X.*

Then for all $p > 0$ and $t_0 > 0$ there is a constant $C > 0$ such that

$$\sup_{t \geq t_0 \varepsilon^{-2}} \mathbb{E}\|u(t)\|^p \leq C\varepsilon^p \tag{2.12}$$

for all sufficiently small $\varepsilon > 0$ and all strong solutions u of (2.3) in X independent of the initial condition. Especially, $\tau_e = \infty$ almost surely for the maximal time of existence of u.

Furthermore, for $q \geq 2$, $\delta > 0$, and $p \in [2, q]$ there is some constant $C > 0$ such *that* $\mathbb{E}\|u(0)\|^q \leq \delta\varepsilon^q$ *for all* $\varepsilon \in (0,1)$ *implies*

$$\sup_{t\geq 0} \mathbb{E}\|u(t)\|^p \leq C\varepsilon^p \quad \text{for all sufficiently small } \varepsilon > 0. \tag{2.13}$$

Additionally, for $t_\varepsilon = \frac{2}{\omega} \ln(\varepsilon^{-1})$ *and all* $p \in [4, q/3]$ *there is a constant* $C > 0$ *such that*

$$\sup_{t\geq t_\varepsilon} \mathbb{E}\|P_s u(t)\|^p \leq C\varepsilon^{3p} \quad \text{for all sufficiently small } \varepsilon > 0. \tag{2.14}$$

The proof is straightforward. But, as it is quite technical, we postpone it to Section 2.3. The main tools are standard a priori type estimates using Itô's formula and Burkholder's inequality.

2.2.2 Residual

With Theorem 2.1 at hand we make the following ansatz

$$u(t) = \varepsilon a(\varepsilon^2 t) + \mathcal{O}(\varepsilon^2) , \quad \text{where } a \in \mathcal{N} .$$

Using a formal calculation completely analogous to the one of Section 1.1.1 yields in lowest order of $\varepsilon > 0$ the following amplitude equation:

$$da = A_c a + \mathcal{F}_c(a) + a d\tilde{\beta} , \tag{2.15}$$

where $\{\tilde{\beta}(T)\}_{T\geq 0}$ defined by $\tilde{\beta}(T) = \varepsilon\beta(\varepsilon^{-2}T)$ is a rescaled version of the Brownian motion β. As usual we consider the equation in the Itô sense. Note again, as explained in Section 1.1.1, that a fixed realization of the amplitude equation obviously depends on ε, but in distribution the solutions are independent of ε.

For a solution a of (2.15) we define the residual

$$\text{Res}(\varepsilon a)(\varepsilon^2 t) = -\varepsilon a(\varepsilon^2 t) + \varepsilon e^{tL} a(0) + \varepsilon^2 \int_0^t e^{(t-\tau)L} a(\varepsilon^2 \tau) d\beta(\tau)$$

$$+ \varepsilon^3 \int_0^t e^{(t-\tau)L} [Aa + \mathcal{F}(a)](\varepsilon^2 \tau) d\tau . \tag{2.16}$$

We show:

Theorem 2.2 **(Residual)** *Let Assumptions 2.1, 2.2, and 2.3 be true. Then for all* $p > \frac{4}{3}$, $\delta > 0$ *and* $T_0 > 0$ *there is a constant* $C > 0$ *such that*

$$P_c \text{Res}(\varepsilon a)(\varepsilon^2 t) = 0$$

and

$$\mathbb{E}\left(\sup_{t\in[0,T_0\varepsilon^{-2}]} \|P_s \text{Res}(\varepsilon a)(\varepsilon^2 t)\|^p\right) \leq C\varepsilon^{3p}$$

for all sufficiently small $\varepsilon > 0$ *and all solutions* a *of (2.15) with* $\mathbb{E}\|a(0)\|^{3p} \leq \delta\varepsilon^{3p}$.

The proof is reasonably easy. It relies only on standard a priori bounds for the solution of the amplitude equation, that will be obtained in Appendix A.2. We can also use higher order corrections, as presented in Section 1.1.1 or [BH04] for additive noise. Then the result is slightly more involved.

Proof. We split the residual into $\mathrm{Res} = P_c\mathrm{Res} + P_s\mathrm{Res}$ and consider both terms separately. First from (2.16) we derive

$$P_c\mathrm{Res}(\varepsilon a)(\varepsilon^2 t) = -\varepsilon a(\varepsilon^2 t) + \varepsilon a(0) + \varepsilon^3 \int_0^t [A_c a + \mathcal{F}_c(a)](\varepsilon^2 \tau)d\tau + \varepsilon^2 \int_0^t a(\varepsilon^2 \tau)d\beta(\tau) .$$

Using the slow time-scale $T = \varepsilon^2 t$ and (2.15) yields immediately

$$P_c\mathrm{Res}(\varepsilon a(\varepsilon^2 t)) = 0 .$$

For $P_s\mathrm{Res}$ we project (2.16) to P_sX:

$$P_s\mathrm{Res}(\varepsilon a)(\varepsilon^2 t) = \varepsilon^3 \int_0^t \mathrm{e}^{(t-\tau)L}[A_s a + \mathcal{F}_s(a)](\varepsilon^2 \tau)d\tau . \qquad (2.17)$$

Using the stability of the semigroup from (2.6) together with Assumption 2.2, we derive

$$\mathbb{E} \sup_{t\in[0,T_0\varepsilon^{-2}]} \|P_s\mathrm{Res}(\varepsilon a)(\varepsilon^2 t)\|^p$$

$$\leq C\varepsilon^{3p} \cdot \mathbb{E} \sup_{t\in[0,T_0/\varepsilon^2]} \left[\int_0^t e^{-(t-\tau)\omega}(1 + (t-\tau))^{-\alpha}\left(\|a(\varepsilon^2\tau)\| + \|a(\varepsilon^2\tau)\|^3 \right)d\tau \right]^p$$

$$\leq C\varepsilon^{3p}(1 + \mathbb{E} \sup_{t\in[0,T_0]} \|a(T)\|^{3p})$$

$$\leq C\varepsilon^{3p} , \qquad (2.18)$$

where we used standard a priori bounds for a. See for example Lemma B.9 for $3p \geq 4$. $\qquad\square$

2.2.3 *Approximation*

Define the remainder R, which is the error of our approximation, as

$$\varepsilon^2 R(t) = u(t) - \varepsilon a(\varepsilon^2 t) . \qquad (2.19)$$

We split

$$R = R_c + R_s \qquad \text{with} \quad R_c = P_cR \text{ and } R_s = P_sR .$$

First we treat R_s using the a priori estimates on $P_s u$. This information on $P_s u$ is not necessary for the result, as we can use cut-off techniques to yield local results, but here it helps to simplify the proofs a lot. The a priori estimates on u are only possible because of the very strong stability assumptions on $\mathcal{F}$. Our main result is the following:

Theorem 2.3 **(Approximation)** *Let Assumptions 2.1, 2.2, and 2.3 be true. For $p > 4$, $T_0 > 0$, and $\delta > 0$ there is a constant $C > 0$ such that for all strong solutions u of (2.3) in X with*

$$\mathbb{E}\|u(0)\|^{3p} \le \delta\varepsilon^{3p} \quad and \quad \mathbb{E}\|P_s u(0)\|^p \le \delta\varepsilon^{3p}$$

for all $\varepsilon \in (0,1)$, we derive

$$\mathbb{E}\left(\sup_{t\in[0,T_0\varepsilon^{-2}]} \|P_s R(t))\|^p \right) \le C\varepsilon^p$$

and

$$\mathbb{E}\left(\sup_{t\in[0,T_0\varepsilon^{-2}]} \|P_c R(t))\|^p \right) \le C$$

for all sufficiently small $\varepsilon > 0$, where a is a solution of (2.15) such that $a(0) = \varepsilon^{-1}P_c u(0)$.

Proof. First we establish in Lemma 2.1 an improvement of the bound obtained in the proof of Theorem 2.1. To be more precise, we obtain bounds on $\mathbb{E}(\sup_{t\in[0,T_0\varepsilon^{-2}]} \|u(t)\|^p)$ and $\mathbb{E}(\sup_{t\in[0,T_0\varepsilon^{-2}]} \|P_s u(t)\|^p)$. Hence, (2.19) implies

$$\mathbb{E}\left(\sup_{t\in[0,T_0\varepsilon^{-2}]} \|P_s R(t)\|^p \right) = \mathbb{E}\left(\sup_{t\in[0,T_0\varepsilon^{-2}]} \|\varepsilon^{-2}P_s u(t)\|^p \right)$$
$$\le C\varepsilon^p . \tag{2.20}$$

We thus proved the bound on R_s. Let us now turn to the bound on R_c. Using (2.9) and (2.16) we derive

$$R(t) = e^{tL}R(0) + \int_0^t e^{(t-\tau)L}[\varepsilon^2 AR + \varepsilon^{-2}(\mathcal{F}(u) - \mathcal{F}(\varepsilon a(\varepsilon^2 \cdot)))](\tau)d\tau$$
$$+ \varepsilon \int_0^t e^{(t-\tau)L}R(\tau)d\beta(\tau) + \varepsilon^{-2}\text{Res}(\varepsilon a)(\varepsilon^2 t) . \tag{2.21}$$

Now

$$\varepsilon^{-2}\Big(\mathcal{F}(u) - \mathcal{F}(\varepsilon a) \Big) = 3\varepsilon^2 \mathcal{F}(a, a, R) + 3\varepsilon^3 \mathcal{F}(a, R, R) + \varepsilon^4 \mathcal{F}(R) .$$

Note that a depends on $\varepsilon^2 t$, while R depends on t. For $R_c = P_c R$ we derive using Theorem 2.2 and $R_c(0) = 0$ (by the definition of $a(0)$)

$$R_c(t) = \int_0^t [\varepsilon^2 A_c R + 3\varepsilon^2 \mathcal{F}_c(a, a, R) + 3\varepsilon^3 \mathcal{F}_c(a, R, R) + \varepsilon^4 \mathcal{F}_c(R)]d\tau$$
$$+ \varepsilon \int_0^t P_c R(\tau)d\beta(\tau) . \tag{2.22}$$

Hence,

$$dR_c(t) = \Big(\varepsilon^2 A_c R_c + 3\varepsilon^2 \mathcal{F}_c(a, a, R_c) + 3\varepsilon^3 \mathcal{F}_c(a, R_c, R_c) + \varepsilon^4 \mathcal{F}_c(R_c) + \varepsilon^2 V_c \Big)dt + \varepsilon R_c d\beta ,$$

where V_c collects all terms at least of order $\mathcal{O}(1)$. For example V_c contains all terms not depending on R_c like, for instance, $\mathcal{F}_c(a, a, R_s)$, $\varepsilon\mathcal{F}_c(a, R_s, R_s)$, or $\varepsilon^2\mathcal{F}_c(R_s)$, together with terms like $\varepsilon\mathcal{F}_c(a, R_c, R)$ or $\varepsilon^2\mathcal{F}_c(R_s, R, R)$, which we can easily bound. To be more precise, it is possible to show

$$\mathbb{E}\|V_c(t)\|^q \le C \quad \text{for} \quad q > \frac{4}{3} \quad \text{and all} \quad t \in [0, T_0\varepsilon^{-2}] .$$

It is straightforward to verify this $\mathcal{O}(1)$-bound on V_c using Hölder's inequality and the following bounds. First we know that $a = \mathcal{O}(1)$ using standard a priori bounds for the amplitude equation. See for example Lemma B.9. Furthermore, $R_s = \mathcal{O}(\varepsilon)$, which was verified in (2.20). Finally $R = \mathcal{O}(\varepsilon^{-1})$ and thus $R_c = \mathcal{O}(\varepsilon^{-1})$, which follows from (2.19), if we use again the bound on a together with the attractivity result of Theorem 2.1 or more specific Lemma 2.1.

Using the dissipativity of $\mathcal{F}_c$ on $\mathcal{N}$ from Assumption 2.2 we derive

$$\langle \mathcal{F}_c(a + \varepsilon R_c) - \mathcal{F}_c(a), R_c \rangle \le 0 .$$

Hence, using this together with Itô's formula and Young's inequality

$$d\|R_c\|^q \le C\varepsilon^2(\|R_c\|^q + \|V_c\|^q)dt + \varepsilon q\|R_c\|^q d\beta . \tag{2.23}$$

Thus,

$$\mathbb{E}\|R_c(t)\|^q \le C\varepsilon^2 \int_0^t \mathbb{E}(\|R_c\|^q + \|V_c\|^q)(\tau)d\tau .$$

Now Gronwall's inequality and the bound on V_c readily imply

$$\sup_{t \in [0, T_0\varepsilon^{-2}]} \mathbb{E}\|R_c(t)\|^q \le C . \tag{2.24}$$

Going back to (2.23) for $q = p/2$ $(p \ge 4)$ yields

$$\sup_{t \in [0, T_0\varepsilon^{-2}]} \|R_c(t)\|^p \le \sup_{t \in [0, T_0\varepsilon^{-2}]} \left(C\varepsilon^2 \int_0^t (\|R_c\|^{p/2} + \|V_c\|^{p/2})(\tau)d\tau \right.$$

$$\left. + C\varepsilon \int_0^t \|R_c(\tau)\|^{p/2} d\beta(\tau) \right)^2$$

$$\le C\varepsilon^4 \left(\int_0^{T_0\varepsilon^{-2}} (\|R_c\|^{p/2} + \|V_c\|^{p/2})(\tau)d\tau \right)^2$$

$$+ C \sup_{t \in [0, T_0\varepsilon^{-2}]} \left(\varepsilon \int_0^t \|R_c(\tau)\|^{p/2} d\beta(\tau) \right)^2 . \tag{2.25}$$

Using Burkholder's inequality (see for example Theorem A.7) and the bound on R_c from (2.24), we easily derive

$$\mathbb{E}\left(\sup_{t \in [0, T_0\varepsilon^{-2}]} \|R_c(t)\|^p \right) \le C .$$

$\square$

2.3 A priori Estimates for u

The following section provides standard a priori estimates for solutions of (2.3). Although they are straightforward, they are nevertheless quite technical. We establish bounds for $\mathbb{E}\|u(t)\|^p$ and $\mathbb{E}\|P_s u(t)\|^p$, which are used in the proof of Theorem 2.1. Furthermore, we bound $\mathbb{E}\sup_{t\in[0,T_0\varepsilon^{-2}]}\|u(t)\|^p$ and in Lemma 2.1 $\mathbb{E}\sup_{t\in[0,T_0\varepsilon^{-2}]}\|P_s u(t)\|^p$. The main idea is to apply Itô's formula to $\|u(t)\|^p$ and to use the strong nonlinear stability condition from (2.7). The main technical obstacle is that a priori we do not know that $\mathbb{E}\|u(t)\|^p$ exists. Therefore we use cut-off techniques.

Proof. **(of Theorem 2.1)** For $p \geq 2$ and $\gamma > 0$ consider smooth bounded $\varphi_{\gamma,p} : [0,\infty) \to \mathbb{R}$ such that $0 \leq \varphi_{\gamma,p}(z) \nearrow \varphi_p(z) = z^{p/2}$. To be more precise, define

$$\varphi_{\gamma,p}(z) := \left(\frac{z}{1+\gamma z}\right)^{p/2} \quad \text{for } z \geq 0. \tag{2.26}$$

It is now easy to check that there are constants C_p and c_p independent of γ such that for $z \geq 0$

$$0 \leq \varphi'_{\gamma,p}(z)z \leq C_p\varphi_{\gamma,p}(z), \qquad -p\varphi_{\gamma,p}(z) \leq \varphi''_{\gamma,p}(z)z^2 \leq C_p\varphi_{\gamma,p}(z),$$

$$\varphi'_{\gamma,p}(z)z^2 = \tfrac{p}{2}\varphi_{\gamma,p}(z)^{(p+2)/p}, \qquad \varphi'_{\gamma,p}(z)z^2 \leq \tfrac{p}{2}\varphi_{\gamma,p-2}(z) = \tfrac{p}{2}\varphi_{\gamma,p}(z)^{(p-2)/p}. \tag{2.27}$$

Apply Itô's formula to $\varphi_{\gamma,p}(\|u(t)\|^2)$ for $t < \tau_e$ to derive

$$
\begin{aligned}
d\varphi_{\gamma,p}(\|u(t)\|^2) &= \varphi'_{\gamma,p}(\|u(t)\|^2)\langle u(t), Lu(t) + \varepsilon^2 Au(t) + \mathcal{F}(u(t))\rangle dt \\
&\quad + \varphi'_{\gamma,p}(\|u(t)\|^2)\|u(t)\|^2[\varepsilon d\beta(t) + \tfrac{1}{2}\varepsilon^2 dt] \\
&\quad + \varphi''_{\gamma,p}(\|u(t)\|^2)\|u(t)\|^4\varepsilon^2 dt.
\end{aligned}
\tag{2.28}
$$

Hence, for $t < \tau_0$ as we are dealing with strong solutions in the sense of Definition 2.4.

$$
\begin{aligned}
&\mathbb{E}\,\varphi_{\gamma,p}(\|u(t)\|^2) - \mathbb{E}\,\varphi_{\gamma,p}(\|u(0)\|^2) \\
&= \int_0^t \mathbb{E}\,\varphi'_{\gamma,p}(\|u(\tau)\|^2)\langle u(\tau), Lu(\tau) + \varepsilon^2 Au(\tau) + \mathcal{F}(u(\tau))\rangle d\tau \\
&\quad + \tfrac{1}{2}\varepsilon^2 \int_0^t \mathbb{E}\,\varphi'_{\gamma,p}(\|u(\tau)\|^2)\|u(\tau)\|^2 d\tau \\
&\quad + \varepsilon^2 \int_0^t \mathbb{E}\,\varphi''_{\gamma,p}(\|u(\tau)\|^2)\|u(\tau)\|^4 d\tau.
\end{aligned}
$$

This implies by (2.10) and (2.27) that $\mathbb{E}\,\varphi_{\gamma,p}(\|u(t)\|^2)$ is differentiable with derivative in $L^1([0,T])$. Thus

$$\partial_t \mathbb{E} \, \varphi_{\gamma,p}(\|u(t)\|^2) = \mathbb{E} \, \varphi'_{\gamma,p}(\|u(t)\|^2)\langle u(t), Lu(t) + \varepsilon^2 Au(t) + \mathcal{F}(u(t))\rangle$$
$$+ \tfrac{1}{2}\varepsilon^2 \mathbb{E} \, \varphi'_{\gamma,p}(\|u(t)\|^2)\|u(t)\|^2 + \varepsilon^2 \mathbb{E} \, \varphi''_{\gamma,p}(\|u(t)\|^2)\|u(t)\|^4 \, .$$

Using Assumption 2.2, (2.27), and Young's inequality we derive

$$\partial_t \mathbb{E} \, \varphi_{\gamma,p}(\|u(t)\|^2) \leq C\varepsilon^4 \mathbb{E} \, \varphi_{\gamma,p-2}(\|u(t)\|^2) - c \, \mathbb{E} \, \varphi_{\gamma,p+2}(\|u(t)\|^2)$$
$$+ \varepsilon^2 \mathbb{E} \, \varphi_{\gamma,p}(\|u(t)\|^2) \, .$$

For any $\delta > 0$ there is a constant $C_\delta > 0$ such that by Young's inequality

$$\varepsilon^2 \mathbb{E} \, \varphi_{\gamma,p}(\|u(t)\|^2) \leq C_\delta \varepsilon^4 \mathbb{E} \, \varphi_{\gamma,p-2}(\|u(t)\|^2) + \delta \mathbb{E} \, \varphi_{\gamma,p+2}(\|u(t)\|^2) \, .$$

Hence,

$$\partial_t \mathbb{E} \, \varphi_{\gamma,p}(\|u(t)\|^2) \leq C\varepsilon^4 \mathbb{E} \, \varphi_{\gamma,p-2}(\|u(t)\|^2) - \frac{c}{2}\mathbb{E} \, \varphi_{\gamma,p}(\|u(t)\|^2)^{(p+2)/p} \, . \qquad (2.29)$$

First for $p = 2$, (2.29) together with well known comparison principles for ODEs (see e.g. Lemma A.7) readily implies

$$\sup_{t \geq t_0\varepsilon^{-2}} \mathbb{E} \, \varphi_{\gamma,2}(\|u(t)\|^2) \leq C\varepsilon^2$$

and hence, using monotone convergence for $\gamma \to 0$,

$$\sup_{t \geq t_0\varepsilon^{-2}} \mathbb{E} \, \|u(t)\|^2 \leq C\varepsilon^2 \, .$$

Thus (2.12) holds for all $p \in (0,2]$.

Suppose in the following that (2.12) holds for some $p - 2$. This together with (2.29) and Hölder's inequality implies

$$\partial_t \mathbb{E} \, \varphi_{\gamma,p}(\|u(t)\|^2) \leq C\varepsilon^{p+2} - C \left(\mathbb{E} \, \varphi_{\gamma,p}(\|u(t)\|^2)\right)^{(p+2)/p} \, . \qquad (2.30)$$

Again we use a comparison principle (cf. Lemma A.7) and monotone convergence for $\gamma \to 0$ to derive

$$\sup_{t \geq t_0\varepsilon^{-2}} \mathbb{E}\|u(t)\|^p \leq C\varepsilon^p \, . \qquad (2.31)$$

This finishes the first part of the proof first for p being a multiple of 2, but then from Hölder's inequality for general p.

The second claim follows from a slight modification of the previous argument, since we know already bounds for the initial condition.

For the last part note that given C_0 it is enough to verify $\mathbb{E}\|P_s u(t_\varepsilon)\|^p \leq C\varepsilon^{3p}$ for all initial conditions $u(0)$ such that $\mathbb{E}\|u(0)\|^p \leq C_0\varepsilon^p$. Using the mild formulation, we can then consider $u(t)$, to be the solution at time t_ε with initial condition $u(t-t_\varepsilon)$, and we know already by (2.13) of Theorem 2.1 that $\mathbb{E}\|u(t)\|^p \leq C_0\varepsilon^p$ uniformly in $t \geq 0$.

We now project (2.9) to $P_s X$ to derive ($u_s = P_s u$)

$$u_s(t) = \mathrm{e}^{tL} u_s(0) + \int_0^t \mathrm{e}^{(t-\tau)L}[\varepsilon^2 A_s u + \mathcal{F}_s(u)](\tau)d\tau + \varepsilon \int_0^t \mathrm{e}^{(t-\tau)L} u_s(\tau)d\beta(\tau) .$$

Hence, by Assumptions 2.1 and 2.2

$$\|u_s(t)\| \le M\mathrm{e}^{-t\omega}\|u(0)\| + \|\varepsilon \int_0^t \mathrm{e}^{(t-\tau)L} u_s(\tau)d\beta(\tau)\|$$

$$+ C \int_0^t (t-\tau)^{-\alpha} \mathrm{e}^{-(t-\tau)\omega}[\varepsilon^2\|u\| + \|u\|^3](\tau)d\tau . \tag{2.32}$$

First, by a standard moment inequality for stochastic integrals (see e.g. Lemma A.5) and Hölder's inequality

$$\mathbb{E}\left\| \int_0^t \mathrm{e}^{(t-\tau)L} u_s(\tau)d\beta(\tau) \right\|^p \le C\mathbb{E}\left(\int_0^t \mathrm{e}^{-(t-\tau)2\omega}\|u_s(\tau)\|^2 d\tau \right)^{p/2}$$

$$\le Ct^{(p-2)/2} \int_0^t \mathrm{e}^{-(t-\tau)p\omega}\mathbb{E}\|u_s(\tau)\|^p d\tau . \tag{2.33}$$

For last term in (2.32) we apply Itô's formula to $\|u(t)\|^{p/2}$ to derive for $p \ge 4$ in a similar way as in (2.28)

$$d\|u(t)\|^{p/2} \le \|u(t)\|^{p/2-2}(C\varepsilon^4 - c\|u\|^4)dt + \frac{p}{2}\varepsilon\|u(t)\|^{p/2}d\beta(t)$$

$$\le C\varepsilon^4\|u(t)\|^{p/2-2} + \frac{p}{2}\varepsilon\|u(t)\|^{p/2}d\beta(t) . \tag{2.34}$$

Hence, for $t \le t_\varepsilon = \frac{2}{\omega}\ln(\varepsilon^{-1})$

$$\|u(t)\|^{p/2} \le \|u(0)\|^{p/2} + C\varepsilon^4 \int_0^{t_\varepsilon} \|u(\tau)\|^{p/2-2}d\tau + C\varepsilon \sup_{t\in[0,t_\varepsilon]} \int_0^t \|u(\tau)\|^{p/2}d\beta(\tau) .$$

Now using $(a+b)^2 \le 2a^2 + 2b^2$

$$\|u(t)\|^p \le C\|u(0)\|^p + C\varepsilon^8 t_\varepsilon \int_0^{t_\varepsilon} \|u(\tau)\|^{p-4}d\tau$$

$$+ C\varepsilon^2 \sup_{t\in[0,t_\varepsilon]} \left(\int_0^t \|u(\tau)\|^{p/2}d\beta(\tau) \right)^2 .$$

Finally using Burkholder's inequality (cf. Theorem A.7) and (2.13) of Theorem 2.1 for $\tilde{p} \le q$

$$\mathbb{E}\left(\sup_{t\in[0,t_\varepsilon]} \|u(t)\|^{\tilde{p}} \right) \le C\varepsilon^{\tilde{p}} + Ct_\varepsilon^2 \varepsilon^{\tilde{p}+4} + C\varepsilon^2 t_\varepsilon \varepsilon^{\tilde{p}}$$

$$\le C\varepsilon^{\tilde{p}} . \tag{2.35}$$

Now using (2.35) for $\tilde{p} = p$ and $\tilde{p} = 3p$

$$\mathbb{E}\left|\int_0^t (t-\tau)^{-\alpha} e^{-(t-\tau)\omega} [\varepsilon^2 \|u\| + \|u\|^3](\tau) d\tau\right|^p \leq C\varepsilon^{3p} \qquad (2.36)$$

for $t \leq t_\varepsilon$. Note that if we control arbitrary moments, then in order to bound (2.36) we do not need bounds on the supremum like the ones in (2.35). We can rely on Hölder's inequality only.

Combining (2.32), (2.33), and (2.36) yields for $t \leq t_\varepsilon$

$$e^{tp\omega} \mathbb{E}\|u_s(t)\|^p \leq C\mathbb{E}\|u(0)\|^p + C\varepsilon^p t_\varepsilon^{(p-2)/2} \int_0^t e^{\tau p\omega} \mathbb{E}\|u_s(\tau)\|^p d\tau + C\varepsilon^{3p} e^{t_\varepsilon \omega p} .$$

And Gronwall's inequality implies for $t \leq t_\varepsilon$

$$\mathbb{E}\|u_s(t)\|^p \leq C\varepsilon^p (1 + \varepsilon^{3p} e^{t_\varepsilon \omega p}) \exp\{C\varepsilon^p t_\varepsilon^{p/2}\} e^{-pt\omega}$$
$$\leq C\varepsilon^p e^{-pt\omega} ,$$

as $e^{t_\varepsilon \omega p} = \varepsilon^{-2p}$. Thus

$$\mathbb{E}\|u_s(t_\varepsilon)\|^p \leq C\varepsilon^{3p} ,$$

and (2.14) follows. $\qquad\qquad\qquad\qquad\qquad\qquad\qquad\qquad\qquad\qquad\quad\square$

The next lemma is a very strong a priori estimate for u. It controls expected values of suprema for time intervals longer than $\mathcal{O}(\ln(\varepsilon^{-1}))$. This improves the results already obtained in the previous proof.

Lemma 2.1   *Let Assumptions 2.1, 2.2, and 2.3 be true.*

   For $p \geq 4$, $T_0 > 0$, and $\delta > 0$ there is a constant $C > 0$ such that for all strong solutions u of (2.3) in X with $\mathbb{E}\|u(0)\|^p \leq \delta\varepsilon^p$ we have

$$\mathbb{E}\left(\sup_{t \in [0, T_0 \varepsilon^{-2}]} \|u(t)\|^p\right) \leq C\varepsilon^p . \qquad (2.37)$$

If additionally $p > 4$ and $\mathbb{E}\|u(0)\|^{3p} \leq \delta\varepsilon^{3p}$, then

$$\mathbb{E}\left(\sup_{t \in [0, T_0 \varepsilon^{-2}]} \|P_s u(t)\|^p\right) \leq C(\varepsilon^{3p} + \mathbb{E}\|P_s u(0)\|^p) . \qquad (2.38)$$

Proof.   Consider (2.34). Burkholder's inequality (cf. Lemma A.3) and (2.13) of Theorem 2.1 yield

$$\mathbb{E}\left(\sup_{t \in [0, T_0 \varepsilon^{-2}]} \varepsilon \int_0^t \|u(\tau)\|^{p/2} d\beta(\tau)\right)^2 \leq C \sup_{t \in [0, T_0 \varepsilon^{-2}]} \mathbb{E}\|u(t)\|^p$$
$$\leq C\varepsilon^p . \qquad (2.39)$$

Furthermore, from (2.34)

$$\mathbb{E}\Big(\sup_{t\in[0,T_0\varepsilon^{-2}]} \|u(t)\|^p \Big)$$

$$\leq \mathbb{E} \sup_{t\in[0,T_0\varepsilon^{-2}]} \Big(\|u(0)\|^{p/2} + \int_0^t C\varepsilon^4 \|u(\tau)\|^{p/2-2}d\tau + \varepsilon \int_0^t \|u(\tau)\|^{p/2}d\beta(\tau) \Big)^2$$

$$\leq C\varepsilon^p + C\varepsilon^8 (T_0\varepsilon^{-2})^2 \cdot \mathbb{E}\Big(\sup_{t\in[0,T_0\varepsilon^{-2}]} \|u(t)\|^{p-4} \Big) ,$$

where we used (2.39) and the condition on $u(0)$. Now by (2.37) the first claim follows easily.

For the second part consider (2.9) and Assumptions 2.1 and 2.2 to derive ($u_s = P_s u$)

$$\|u_s(t)\| \leq Me^{-t\omega}\|u_s(0)\| + \|\varepsilon \int_0^t e^{(t-\tau)L}u_s(\tau)d\beta(\tau)\|$$

$$+ C\int_0^t (t-\tau)^{-\alpha}e^{-(t-\tau)\omega}[\varepsilon^2\|u\| + \|u\|^3](\tau)d\tau .$$

Now using a maximal inequality for a stochastic integral of convolution type (See e.g. Lemma A.4) yields

$$\mathbb{E} \sup_{t\in[0,T_0\varepsilon^{-2}]} \|u_s(t)\|^p \leq C\mathbb{E}\|u_s(0)\|^p + C\varepsilon^{p-2} \cdot \sup_{t\in[0,T_0\varepsilon^{-2}]} \mathbb{E}\|u_s(t)\|^p$$

$$+ C \cdot \mathbb{E}\Big(\sup_{t\in[0,T_0\varepsilon^{-2}]} [\varepsilon^2\|u(t)\| + \|u(t)\|^3]^p \Big)$$

$$\leq C \cdot \mathbb{E}\|u_s(0)\|^p + C\varepsilon^{3p} + C\varepsilon^{p-2}\mathbb{E}\Big(\sup_{t\in[0,T_0\varepsilon^{-2}]} \|u_s(t)\|^p \Big) ,$$

where we used (2.37) for p and $3p$. Now for $\varepsilon > 0$ sufficiently small we finish the proof. □

2.4 Results for Quadratic Nonlinearities

This section states rigorous results for the approximation via amplitude equations for quadratic nonlinearities. We focus only on the interesting case, where $P_cB(a,a) = 0$, which was discussed for additive noise on a formal level in Section 1.1.3. The case with $P_cB(a,a) \neq 0$ is similar to the cubic case. The formal result for our case is completely analogous to the one stated in Section 1.1.2, we summarise details below. Nevertheless, in this case in general we cannot bound moments of solutions. We have to use cut-off techniques in order to use moments.

Here we present a somewhat simpler model with multiplicative noise, in order to simplify the presentation. We review the results of [Blö05a] for additive noise in

Section 2.6. In [Blö05a] also fractional (i.e. smoother) additive noise was used, but we do not focus on that.

Consider

$$\partial_t u = Lu + \varepsilon^2 Au + B(u, u) + \varepsilon u\dot{\beta} \,, \tag{2.40}$$

with L and A as in Assumption 2.1 and 2.2, and B some bilinear mapping defined later on in Assumption 2.4.

Let us recall the formal derivation of the amplitude equation, which is similar to Section 1.1.3. Plugging the ansatz

$$u(t) = \varepsilon a(\varepsilon^2 t) + \varepsilon^2 \psi_o(\varepsilon^2 t)$$

with $a \in \mathcal{N}$ and $\psi_o \in P_s X$ into (2.40), we derive in lowest order of $\varepsilon > 0$

$$\mathcal{O}(\varepsilon^2) \text{ in } \mathcal{N}: \qquad 0 = B_c(a, a) \,, \tag{2.41}$$

$$\mathcal{O}(\varepsilon^3) \text{ in } \mathcal{N}: \qquad \partial_T a = A_c a + 2B_c(a, \psi_o) + a\partial_T \tilde{\beta} \,, \tag{2.42}$$

$$\mathcal{O}(\varepsilon^2) \text{ in } P_s X: \qquad 0 = L\psi_o + B_s(a, a) \,. \tag{2.43}$$

Note that $\tilde{\beta}(T) = \varepsilon\beta(T\varepsilon^{-2})$ is again a rescaled Brownian motion. From (2.41) we see that $B_c(a, a) = 0$ ($B_c := P_c B$, as usual) is necessary for the approach presented. Finally, projecting (2.43) to $P_s X$ and solving for ψ_o yields

$$\partial_T a = A_c a - 2B_c(a, L_s^{-1} B_s(a, a)) + a\partial_T \tilde{\beta} \,, \tag{2.44}$$

or in integrated form

$$a(T) = a(0) + \int_0^T [A_c a - 2B_c(a, L_s^{-1} B_s(a, a))](\tau)d\tau + \int_0^T a(\tau)d\tilde{\beta}(\tau) \,, \tag{2.45}$$

where we consider as before Itô-differentials. Nevertheless, as discussed before in Section 2.1, we could also consider Stratonovič-differentials everywhere, and still obtain the same result. An interesting feature of (2.45) is that the amplitude equation involves a cubic nonlinearity. Therefore, we can expect nonlinear stability of the amplitude equation, which is in general not present for the SPDE.

For the rigorous approximation result we slightly modify the correction ψ_o, by adding higher order terms. The first order approximation remains unchanged.

Let a be a $\mathcal{N}$-valued solution of (2.45). For the second order correction define for some $\psi(0) \in P_s X$

$$\psi(T) = -L_s^{-1} B_s(a(T), a(T)) + e^{T\varepsilon^{-2}L}(L_s^{-1} B_s(a(0), a(0)) + \psi(0)) \,, \tag{2.46}$$

and thus the approximation is

$$\varepsilon w(t) = \varepsilon a(\varepsilon^2 t) + \varepsilon^2 \psi(\varepsilon^2 t) \,. \tag{2.47}$$

This approximation is different from the one derived by the formal calculation. But we can use the formal argument that

$$e^{T\varepsilon^{-2}L}P_s \approx -\varepsilon^2 L_s^{-1}\delta(T) \qquad (2.48)$$

with δ being the Delta-distribution. Equation (2.48) is motivated by

$$\lambda e^{-t\lambda} \to \delta(t) \quad \text{for} \quad \lambda \to \infty .$$

Or, to be more precise,

$$\int_0^\infty \lambda e^{-t\lambda}\varphi(t)dt \to \varphi(0) = \int_0^\infty \varphi(t)\delta(t)dt \quad \text{for} \quad \lambda \to \infty$$

for all bounded continuous functions φ.

With (2.48) we see formally that we expect to have added only higher order term to the original second order correction

$$\psi_o(T) = -L_s^{-1}B_s(a(T), a(T)) . \qquad (2.49)$$

We see later in the proof that these higher order terms are localised at time $T = 0$ and will improve bounds for small times. Note that they decay exponentially fast in time.

If we would use $\varepsilon a(\varepsilon^2 t) + \varepsilon^2 \psi_o(\varepsilon^2 t)$ as our approximation, then the approximation will have a large error at $t = 0$. Let us indicate this at $t = 0$. The error $u(0) - \varepsilon a(0) + \varepsilon^2 L_s^{-1}B_s(a(0), a(0))$ may just be $\mathcal{O}(\varepsilon^2)$. In contrast to that using (2.47) leaves us with an error $u(0) - \varepsilon w(0) = 0$, if we choose $a(0)$ and $\psi(0)$ appropriately.

Let us comment furthermore on the role of ψ or ψ_o in the proofs. Neither of them has a big impact on the final result, as we can only verify that $u(t) = \varepsilon w(t) + \mathcal{O}(\varepsilon^2)$. The main reason is that we ignored the terms of order $\mathcal{O}(\varepsilon^2)$ in $P_c u(t)$, that are actually present. The correction $\varepsilon^2 \psi$ therefore simplifies the proof.

Let us give another equivalent approach. Instead of (2.46) we could use

$$\tilde{\psi}(T) = e^{TL\varepsilon^{-2}}\psi(0) + \varepsilon^{-2}\int_0^T e^{(T-\tau)L\varepsilon^{-2}}B_s(a(\tau))d\tau . \qquad (2.50)$$

Using Lemma 2.2, one can actually show that ψ and $\tilde{\psi}$ differ only in higher order terms. And again a formal Delta-distribution estimate of the type $\varepsilon^{-2}e^{(T-\tau)\varepsilon^{-2}L}P_s \approx -L_s^{-1}\delta(T - \tau)$, shows on a formal level that we recover (2.49) from (2.50) in the limit $\varepsilon \to 0$.

Let us fix our assumptions for B and A.

Assumption 2.4   Let $A, B : D(L) \to X$ such that $A \in \mathcal{L}(X, X^{-\alpha})$ for some $\alpha \in [0, 1)$. Assume furthermore that we can extend B to a continuous bilinear operator $B \in \mathcal{L}_2(X, X^{-\alpha})$. Suppose that $B_c(a, b) = 0$ for all $a, b \in \mathcal{N}$, where as usual $B_c = P_c B$ and $B_s = P_s B$ for short. We also abbreviate the quadratic form $B(a) := B(a, a)$.

Furthermore let the following strong stability condition for the amplitude equation (2.44) or (2.45) be true.

$$\langle B_c(a, L_s^{-1} B_s(a, a)), a \rangle > 0 \quad \text{for all } a \in \mathcal{N}, \ a \neq 0 .$$

We consider only mild solutions of (2.40) given by the following definition.

Definition 2.5   Let Assumptions 2.1, 2.3, and 2.4 be true. We call an X-valued stochastic process $\{u(t)\}_{t\geq 0}$ a *mild solution* of (2.40), if it is adapted to the filtration $\{\mathcal{F}_t\}_{t\geq 0}$ and if there is a positive stopping time τ^* such that $u \in C^0([0, \tau^*), X)$ and the following variation of constants formula

$$u(t) = e^{tL} u(0) + \varepsilon \int_0^t e^{(t-\tau)L} u(\tau) d\beta(\tau) + \int_0^t e^{(t-\tau)L} [\varepsilon^2 Au + B(u)](\tau) d\tau \quad (2.51)$$

holds for all $t \in [0, \tau^*)$.

Remark 2.4   *For convenience, we will always assume that solutions in the sense of Definition 2.5 exist. Usually, the existence of unique mild solutions in X is standard under our assumptions. We comment for instance on the case of additive noise in more detail after Assumption 2.10. For a general treatment see for example [DPZ92]. Again the adaptedness of a mild solution to the filtration of the Brownian motion β follows usually directly from (2.51).*

Due to the quadratic nonlinearity we usually expect only local solutions to exist. This means there is in general no lower bound $T > 0$ such that $\tau^* > T$. Nevertheless in some models it is possible to verify global existence of solutions (i.e., $\tau^* = \infty$), but there are many open problems about the existence of unique global mild solutions. One example is the celebrated Navier-Stokes problem.

2.4.1   *Attractivity*

We use a cut-off technique, as in general we cannot control moments of solutions. There are some special cases like for instance one-dimensional Burgers, surface growth, or Kuramoto-Sivashinsky equation (see [BGR02; DPDT94; DE01]), where we actually can derive bounds for moments. But for our results it is enough to cut off the nonlinearity for large solutions, in order to keep it small for solutions that get too large.

This technique is well known for SDEs with blow-ups. See for example [McK69]. For a detailed discussion see Section 6.3 of [HT94]. The idea is always to cut off the nonlinearities, in order to derive bounds for moments and to compute probabilities. But solutions of the modified equation with cut-off and the original equation coincide, as long as both are small. Note that for the local attractivity result we are anyway only interested in solutions that are small. To be more precise we look at solutions of order $\mathcal{O}(\varepsilon)$.

The main result is a local attractivity result for solutions of order $\mathcal{O}(\varepsilon)$. It shows that if $u(0)$ is of order $\mathcal{O}(\varepsilon)$, then at some time $t_\varepsilon = \mathcal{O}(\ln(\varepsilon^{-1}))$ the probability is almost 1 that $u(t_\varepsilon)$ is still of order $\mathcal{O}(\varepsilon)$, but $P_s u(t_\varepsilon)$ decreased to order $\mathcal{O}(\varepsilon^2)$.

Theorem 2.4 (Attractivity) *Let Assumptions 2.1, 2.3, and 2.4 be true.*
For all small $\kappa > 0$, all $\delta > 0$ and $p > 0$ there are constants $C > 0$, $\delta_1, \delta_2 > 0$ such that for $t_\varepsilon = \frac{2}{\omega} \ln(\varepsilon^{-1})$ and all mild solutions in the sense of Definition 2.5

$$\mathbb{P}\Big(\|u(t_\varepsilon)\| \le \delta_1 \varepsilon, \ \|P_s u(t_\varepsilon)\| \le \delta_2 \varepsilon^2\Big) \ge \mathbb{P}\Big(\|u(0)\| \le \delta\varepsilon\Big) - C\varepsilon^p$$

for all $\varepsilon \in (0,1)$.

The proof relies on the linear stability of (2.40) and cut-off techniques. We postpone the proof to Section 2.4.4 as it is not difficult but technical. Let us first discuss the results for the residual and the approximation.

2.4.2 *Residual*

For a solution of the amplitude equation (2.45) and some $\psi(0)$, we consider the approximation εw given by (2.47). The residual of εw is as usual defined as

$$\text{Res}(\varepsilon w)(t) = -\varepsilon w(t) + \varepsilon e^{tL} w(0) + \varepsilon^2 \int_0^t e^{(t-\tau)L} w(\tau) d\beta(\tau)$$

$$+ \int_0^t e^{(t-\tau)L} [\varepsilon^3 A w + \varepsilon^2 B(w)](\tau) d\tau . \tag{2.52}$$

Theorem 2.5 (Residual) *Let Assumptions 2.1, 2.3, and 2.4 be true.*
For $p > 4$, $\delta > 0$, $T_0 > 0$ there is a constant $C > 0$ such that for all approximations defined by (2.46) and (2.47), where a is a solution of (2.45), with $\mathbb{E}\|a(0)\|^{4p} \le \delta$ and $\mathbb{E}\|\psi(0)\|^{2p} \le \delta$ we have

$$\mathbb{E}\Big(\sup_{t \in [0, T_0 \varepsilon^{-2}]} \|P_c \text{Res}(\varepsilon w)(t)\|^p\Big) \le C\varepsilon^{2p} \tag{2.53}$$

and

$$\mathbb{E}\Big(\sup_{t \in [0, T_0 \varepsilon^{-2}]} \|P_s \text{Res}(\varepsilon w)(t)\|^p\Big) \le C\varepsilon^{3p-2} .$$

Furthermore $P_c \text{Res}(\varepsilon w)$ is differentiable with

$$\partial_t P_c \text{Res}(\varepsilon w)(t) = \varepsilon^4 [A_c \psi + B_c(\psi)](\varepsilon^2 t) .$$

The proof below is straightforward using the key Lemma 2.2. This lemma is a purely technical estimate, and we postpone the proof to Section 2.4.4.

Proof.   From (2.52) and (2.47)

$$P_c \mathrm{Res}(\varepsilon w)(t) = -\varepsilon a(\varepsilon^2 t) + \varepsilon a(0) \tag{2.54}$$

$$+\varepsilon^3 \int_0^t [A_c a + 2B_c(a, \psi)](\varepsilon^2 \tau) d\tau + \varepsilon^2 \int_0^t a(\varepsilon^2 \tau) d\beta(\tau) \tag{2.55}$$

$$+\varepsilon^4 \int_0^t [A_c \psi + B_c(\psi)](\varepsilon^2 \tau) d\tau .$$

Using a substitution and (2.45) the terms in (2.54) and (2.55) cancel. Thus,

$$P_c \mathrm{Res}(\varepsilon w)(t) = \varepsilon^4 \int_0^t [A_c \psi + B_c(\psi)](\varepsilon^2 \tau) d\tau . \tag{2.56}$$

For the remaining terms in (2.56) we use the definition (2.46) of ψ together with the fact that $A_c \in \mathcal{L}(X, X)$, $B_c \in \mathcal{L}_2(X, X)$, and $L_s^{-1} \in \mathcal{L}(X^{-\alpha}, X)$. We immediately see that all terms are of order $\mathcal{O}(\varepsilon^2)$, if we use Lemma B.9 to bound the $4p$-th moment of a for $p \geq 1$. Hence, (2.53) follows easily. Furthermore, the derivative of $P_c \mathrm{Res}$ is obvious.

The estimate for $P_s \mathrm{Res}$ is more involved:

$$P_s \mathrm{Res}(\varepsilon w)(t) = \varepsilon^3 \int_0^t \mathrm{e}^{(t-\tau)L} \psi(\varepsilon^2 \tau) d\beta(\tau)$$

$$-\varepsilon^2 \psi(t\varepsilon^2) + \varepsilon^2 \mathrm{e}^{tL} \psi(0) + \varepsilon^2 \int_0^t \mathrm{e}^{(t-\tau)L} B_s(a(\varepsilon^2 \tau)) d\tau \tag{2.57}$$

$$+\varepsilon^3 \int_0^t \mathrm{e}^{(t-\tau)L} [A_s(a + \varepsilon \psi) + 2B_s(a, \psi) + \varepsilon B_s(\psi)](\varepsilon^2 \tau) d\tau . \tag{2.58}$$

The terms in (2.57) are bounded due to some straightforward estimates and Lemma 2.2 (needs $p > 4$), which we also postpone to Section 2.4.4. Recall (2.46) for the definition of $\psi(T)$. Now Lemma 2.2 basically states that

$$\mathbb{E} \sup_{t \in [0, T_0 \varepsilon^{-2}]} \left\| -\varepsilon^2 \psi(t\varepsilon^2) + \varepsilon^2 \mathrm{e}^{tL} \psi(0) + \varepsilon^2 \int_0^t \mathrm{e}^{(t-\tau)L} B_s(a(\varepsilon^2 \tau)) d\tau \right\|^p \leq C\varepsilon^{3p-2} .$$

The terms in (2.58) are bounded by using (2.6), Assumption 2.4, the bound on $\psi(0)$ and standard a priori estimates for $a = \mathcal{O}(1)$ (cf. Lemma B.9) and hence ψ :

$$\mathbb{E} \sup_{t \in [0, T_0 \varepsilon^{-2}]} \left\| \varepsilon^3 \int_0^t \mathrm{e}^{(t-\tau)L} [A_s(a + \varepsilon \psi) + 2B_s(a, \psi) + \varepsilon B_s(\psi)](\varepsilon^2 \tau) d\tau \right\|^p \leq C\varepsilon^{3p} ,$$

where we used that due to (2.6) for $t \leq T_0 \varepsilon^{-2}$

$$\left\| \int_0^t \mathrm{e}^{(t-\tau)L} f(\tau) d\tau \right\| \leq M_\alpha \cdot \int_0^t (1 + \tau^{-\alpha}) \mathrm{e}^{-\tilde{\omega} \tau} d\tau \cdot \sup_{\tau \in [0, T_0/\varepsilon^2]} \|f(\tau)\|_{X^{-\alpha}} .$$

The remaining term is bounded by Theorem A.7 and the bounds on a and ψ. $\square$

2.4.3 *Approximation*

For a solution u of (2.51) define

$$a(0) = \varepsilon^{-1} P_c u(0) \quad \text{and} \quad \psi(0) = \varepsilon^{-2} P_s u(0) . \tag{2.59}$$

Now let a be a solution of (2.45) with initial condition $a(0)$ and define w and ψ as in (2.47) and (2.46). Then we can show that $u(t) \approx \varepsilon w(t)$ in the following sense:

Theorem 2.6 (Approximation) *Let Assumptions 2.1, 2.3, and 2.4 be true.*

For $\delta_1, \delta_2 > 0$, $\tilde{\kappa} \in (0,1]$, and $T_0 > 0$ there is some $\eta > 0$ and some constant $C > 0$ such that for all solutions u of (2.51) and approximations a and ψ defined by (2.59), (2.47), and (2.46) we have

$$\mathbb{P}\left(\sup_{t \in [0, T_0 \varepsilon^{-2}]} \|u(t) - \varepsilon w(t)\| \leq C\varepsilon^{2-\tilde{\kappa}} \right)$$

$$\geq 1 - 2\mathbb{P}\left(\|P_s u(0)\| > \delta_2 \varepsilon^2 \right) - 2\mathbb{P}\left(\|P_c u(0)\| > \delta_1 \varepsilon \right) - C\varepsilon^{\eta}$$

for all $\varepsilon \in (0,1)$.

The proof will be given in the next section. Let us first comment on the improvements of the result compared to older results.

Remark 2.5 *In the proof we need $\|a(t)\|^2$ to be bounded uniformly in $t \in [0, T_0]$ by $\gamma \ln(\varepsilon^{-1})$ for some small $\gamma > 0$ (cf. (2.76)). Hence, as we rely on Theorem B.9 we cannot improve the result to large $\eta > 0$ (cf. equation (2.79)). The main obstacle is that we can only bound certain exponential moments of $\|a\|^2$ and not of higher powers. Nevertheless, this is an improvement to the results of [Blö05a], where the probability was just small without any order in ε. In principle it is easy to thoroughly compute all constants, in order to provide a uniform lower bound on η independent of other constants like δ_j, T_0, and κ. But, as we expect the bound not to be large, for simplicity of presentation we do not focus on that.*

Remark 2.6 *For special types of nonlinearities we will see from the proof that it is possible to improve the result of Theorem 2.6 significantly. We need information about the sign of certain multi-linear functionals. The first one is of the type $F_1(a, a, R_c, R_c) = \langle B_c(\psi, R_c), R_c \rangle$, while the second is given by $F_2(a, a, R_c, R_c) = \langle B_c(a, R_s), R_c \rangle + "Error"$. Recall that ψ depends quadratically on a, and we will see later in the proof that R_s is a function of R_c. Thus a statement of these results is quite technical, but sometimes easy to check for given B and L. The improvement is that we can use standard a priori type estimates for (2.70) and (2.71), where all the critical terms responsible for the bad order in the proof of Theorem 2.6 disappear.*

2.4.4 *Proofs*

This section gives the postponed proofs for Theorem 2.4, Lemma 2.2, and Theorem 2.6. We first provide the proof of the attractivity result.

Proof.   **(of Theorem 2.4)** The main ingredients of the proof are a cut-off technique and the linear stability of L. Fix some $\rho \in C^\infty(\mathbb{R})$ such that $\rho(x) = 1$ for $x \leq 1$ and $\rho(x) = 0$ for $x \geq 2$. Define for some small $\kappa > 0$

$$A^{(\rho)}(u) = \rho(\|u\|\varepsilon^{-1+\kappa}) \cdot Au \quad \text{and} \quad B^{(\rho)}(u) = \rho(\|u\|\varepsilon^{-1+\kappa}) \cdot B(u) .$$

Moreover, define

$$u^{(\rho)}(0) = \begin{cases} u(0) : \text{for } \|u(0)\| \leq \delta\varepsilon \\ 0 : \text{otherwise.} \end{cases}$$

Let $u^{(\rho)}$ be the solution of (2.51) with $A^{(\rho)}$ and $B^{(\rho)}$ instead of A and B and initial condition $u^{(\rho)}(0)$. Thus,

$$u^{(\rho)}(t) = \mathrm{e}^{tL}u^{(\rho)}(0) + \varepsilon \int_0^t \mathrm{e}^{(t-\tau)L}u^{(\rho)}(\tau)d\beta(\tau)$$
$$+ \int_0^t \mathrm{e}^{(t-\tau)L}\left[\varepsilon^2 A^{(\rho)}(u^{(\rho)}) + B^{(\rho)}(u^{(\rho)})\right](\tau)d\tau . \qquad (2.60)$$

The existence of a unique solution $u^{(\rho)}$ is standard (cf. [DPZ92]), as we have global Lipschitz nonlinearities.

Define furthermore the stopping time

$$\tau_\rho = \begin{cases} \inf\{t > 0 : \|u^{(\rho)}(t)\| > \varepsilon^{1-\kappa}\} : \text{for } \|u(0)\| \leq \delta\varepsilon \\ 0 \qquad\qquad\qquad\qquad\qquad : \text{otherwise.} \end{cases}$$

Obviously, $u(t) = u^{(\rho)}(t)$ for $0 \leq t \leq \tau_\rho$.

Let us first establish bounds on $u^{(\rho)}$. From (2.60), we derive using the definition of $u^{(\rho)}(0)$, $A^{(\rho)}$, $B^{(\rho)}$, and the assumptions on e^{tL} that

$$\|u^{(\rho)}(t)\| \leq C\delta\varepsilon + Ct^{1-\alpha}(\varepsilon^{3-\kappa} + \varepsilon^{2-2\kappa}) + \varepsilon\left\|\int_0^t \mathrm{e}^{(t-\tau)L}u^{(\rho)}(\tau)d\beta(\tau)\right\| . \qquad (2.61)$$

Hence, using Hölder's and a moment inequality for stochastic integrals (cf. Lemma A.5) we obtain for $q > 2$ and $t \leq t_\varepsilon$

$$\mathbb{E}\left\|\int_0^t \mathrm{e}^{(t-\tau)L}u^{(\rho)}(\tau)d\beta(\tau)\right\| \leq C\mathbb{E}\int_0^t \|u^{(\rho)}(\tau)\|^q d\tau \cdot t_\varepsilon^{q/(q-2)} .$$

With Gronwall's inequality, we easily derive

$$\mathbb{E}\|u^{(\rho)}(t)\|^q \leq C\varepsilon^q \quad \text{for all} \quad t \leq t_\varepsilon = \frac{2}{\omega}\ln(\varepsilon^{-1}) . \qquad (2.62)$$

Now $\mathbb{P}(\|u^{(\rho)}(t_\varepsilon)\| > \varepsilon^{1-\kappa}) \leq C\varepsilon^q$ follows immediately, but using (2.61) it is easy to improve this to

$$\mathbb{P}\left(\|u^{(\rho)}(t_\varepsilon)\| \leq \delta_1\varepsilon\right) \geq \mathbb{P}\left(\|\int_0^{t_\varepsilon} \mathrm{e}^{(t-\tau)L}u^{(\rho)}(\tau)d\beta(\tau)\| \leq 1\right) \geq 1 - C\varepsilon^p \qquad (2.63)$$

for some large constant $\delta_1 > 0$ and all $p > 0$, where we used Chebychev's inequality, Lemma A.5, and (2.62). Note that the constant C depends obviously on p.

Let us now turn to a uniform bound in time t. From (2.61) we derive

$$\|u^{(\rho)}(t)\| \leq C\varepsilon + \varepsilon \cdot \left\| \int_0^t e^{(t-\tau)L} u^{(\rho)}(\tau)d\beta(\tau) \right\| \quad \text{for all } t \in [0, t_\varepsilon] .$$

First, using a maximal inequality for stochastic integrals of convolution type (cf. Lemma A.4) together with (2.62), we obtain for all $q > 4$

$$\mathbb{E}\left(\sup_{t\in[0,t_\varepsilon]} \|u^{(\rho)}(t)\|^q \right) \leq C\varepsilon^q + C\varepsilon^q \cdot \mathbb{E} \sup_{t\in[0,t_\varepsilon]} \left\| \int_0^t e^{(t-\tau)L} u^{(\rho)}(\tau)d\beta(\tau) \right\|^q$$

$$\leq C\varepsilon^q . \tag{2.64}$$

Secondly, using again the maximal inequality from Lemma A.4 for some constant $\delta_3 > 0$

$$\mathbb{P}\left(\sup_{t\in[0,t_\varepsilon]} \|u^{(\rho)}(t)\| \leq \delta_3\varepsilon \right) \geq \mathbb{P}\left(\sup_{t\in[0,t_\varepsilon]} \left\| \int_0^{t_\varepsilon} e^{(t-\tau)L} u^{(\rho)}(\tau)d\beta(\tau)\right\| \leq 1 \right)$$

$$\geq 1 - C\varepsilon^p . \tag{2.65}$$

For $u_s^{(\rho)} = P_s u^{(\rho)}$ we first derive from (2.60), by using $e^{-t_\varepsilon \omega/2} = \varepsilon$ and $\|u^{(\rho)}(0)\| \leq \delta\varepsilon$, that for $t \in [t_\varepsilon/2, t_\varepsilon]$

$$\|u_s^{(\rho)}(t)\| \leq C\varepsilon^2 + C \sup_{t\in[0,t_\varepsilon]} \|u^{(\rho)}(t)\|^2 + \varepsilon \cdot \left\| \int_0^t e^{(t-\tau)L} u_s^{(\rho)}(\tau)d\beta(\tau) \right\| .$$

Combining all estimates it is easy to show that

$$\mathbb{E}\|u_s^{(\rho)}(t)\|^q \leq C\varepsilon^{2q} \quad \text{for all} \quad t \in [t_\varepsilon/2, t_\varepsilon] . \tag{2.66}$$

Moreover, using the mild formulation with initial time $t_\varepsilon/2$, we derive

$$\|u_s^{(\rho)}(t_\varepsilon)\| \leq C\varepsilon^3 + C \sup_{t\in[t_\varepsilon/2,t_\varepsilon]} \|u^{(\rho)}(t)\|^2 + \varepsilon \cdot \left\| \int_{t_\varepsilon/2}^{t_\varepsilon} e^{(t_\varepsilon-\tau)L} u_s^{(\rho)}(\tau)d\beta(\tau) \right\| .$$

Hence, using (2.65), the maximal inequality of Lemma A.4, and (2.66), it is straightforward to show that

$$\mathbb{P}\left(\|u_s^{(\rho)}(t_\varepsilon)\| \leq \delta_2\varepsilon^2 \right) \geq 1 - C\varepsilon^p \tag{2.67}$$

for some suitable constant $\delta_2 > 0$.

Let us now carry over the result to u. By the definition of the stopping time τ_ρ and Chebychev's inequality

$$
\begin{aligned}
\mathbb{P}(\tau_\rho \geq t_\varepsilon) &= \mathbb{P}\Big(\|u(0)\| \leq \delta\varepsilon,\ \sup_{t\in[0,t_\varepsilon]} \|u^{(\rho)}(t)\| \leq \varepsilon^{1-\kappa}\Big) \\
&\geq \mathbb{P}\Big(\|u(0)\| \leq \delta\varepsilon\Big) - \mathbb{P}\Big(\sup_{t\in[0,t_\varepsilon]} \|u^{(\rho)}(t)\| > \varepsilon^{1-\kappa}\Big) \\
&\stackrel{(2.64)}{\geq} \mathbb{P}\Big(\|u(0)\| \leq \delta\varepsilon\Big) - C\varepsilon^p
\end{aligned}
\tag{2.68}
$$

for all $p > 0$, where the constant depends on p and δ. Moreover,

$$
\begin{aligned}
&\mathbb{P}\Big(\|u(t_\varepsilon)\| \leq \delta_1\varepsilon,\ \|P_s u(t_\varepsilon)\| \leq \delta_2\varepsilon^2\Big) \\
&\quad \geq \mathbb{P}\Big(\tau_\rho \geq t_\varepsilon,\ \|u^{(\rho)}(t_\varepsilon)\| \leq \delta_1\varepsilon,\ \|P_s u^{(\rho)}(t_\varepsilon)\| \leq \delta_2\varepsilon^2\Big) \\
&\quad \stackrel{(2.68)}{\geq} \mathbb{P}\Big(\|u(0)\| \leq \delta\varepsilon\Big) - C\varepsilon^p \\
&\qquad - \mathbb{P}\Big(\|u^{(\rho)}(t_\varepsilon)\| > \delta_1\varepsilon\Big) - \mathbb{P}\Big(\|P_s u^{(\rho)}(t_\varepsilon)\| > \delta_2\varepsilon^2\Big).
\end{aligned}
$$

Using (2.63) and (2.67), Theorem 2.4 is proved.   □

We need the following key lemma to bound the residual.

Lemma 2.2   *For ψ defined in (2.46) and L from Assumption 2.1. For all $p > 4$, $\delta > 0$, and $T_0 > 0$ there is a constant $C > 0$ such that*

$$
\mathbb{E} \sup_{t\in[0,T_0\varepsilon^{-2}]} \Big\| e^{tL}\psi(0) - L\int_0^t e^{(t-\tau)L}\psi(\varepsilon^2\tau)d\tau - \psi(\varepsilon^2 t)\Big\|^p \leq C\varepsilon^{p-2}
$$

for all solutions a of (2.45) such that $\mathbb{E}\|a(0)\|^{4p} \leq \delta$ and all $\psi(0)$ such that $\mathbb{E}\|\psi(0)\|^p \leq \delta$.

Proof.   For smooth f we obtain via integration by parts

$$
e^{tL}f(0) - L\int_0^t e^{(t-\tau)L}f(\varepsilon^2\tau)d\tau - f(\varepsilon^2 t) = \int_0^T e^{(T-\tau)L\varepsilon^{-2}}df(\tau)
$$

with $T = \varepsilon^2 t$, where the differential df is a Riemann-Stieltjes integral. For the stochastic process we use Itô's formula

$$
d(-L_s^{-1}B_s(a,a)) = -2L_s^{-1}B_s(a,da) - L_s^{-1}B_s(da,da)
$$

and the amplitude equation (2.44) to derive

$$
\begin{aligned}
d\psi &= -L_s^{-1}B_s(a, a + 2A_c a)\, dT + 4L_s^{-1}B_s\Big(a, B_c(a, L_s^{-1}B_s(a,a))\Big)dT \\
&\quad -2L_s^{-1}B_s(a,a)d\tilde\beta + \varepsilon^{-2}Le^{TL\varepsilon^{-2}}\big(L_s^{-1}B_s(a(0),a(0)) + \psi(0)\big)\, dT \\
&=: G_1 dT + G_2 d\tilde\beta + G_3 dT
\end{aligned}
\tag{2.69}
$$

with $G_j : [0, T_0] \to P_s X$, where

$$\|G_1\| \le C(\|a\|^4 + 1) \qquad \text{and} \qquad \|G_2\| \le C\|a\|^2 \, ,$$

which follows from the properties of A and B and the fact that $L_s^{-1} B_s : X^2 \to X$ is bilinear and continuous by Assumption 2.4.

Now we easily derive that

$$e^{tL}\psi(0) - L \int_0^t e^{(t-\tau)L}\psi(\varepsilon^2\tau)d\tau - \psi(\varepsilon^2 t)$$
$$= \int_0^T e^{(T-\tau)L\varepsilon^{-2}} \left(G_1(\tau)d\tau + G_2(\tau)d\tilde{\beta}(\tau) + G_3(\tau)d\tau \right) \, .$$

First,

$$\int_0^T e^{(T-\tau)L\varepsilon^{-2}} G_3(\tau)d\tau = \varepsilon^{-2} L T e^{TL\varepsilon^{-2}}(L_s^{-1}B_s(a(0), a(0)) + \psi(0)) \, ,$$

which is easily bounded in p-th mean, using that $\{tLe^{tL}\}_{t>0}$ is a uniformly bounded family of operators (cf. (2.5)), and using the bound on $a(0)$ and $\psi(0)$.

Secondly,

$$\left\| \int_0^T e^{(T-\tau)L\varepsilon^{-2}} G_1(\tau)d\tau \right\| \le C \int_0^T e^{-\tau\omega\varepsilon^{-2}} d\tau \cdot \left(\sup_{T\in[0,T_0]} \|a(T)\|^4 + 1 \right)$$
$$= C\varepsilon^2 \left(\sup_{T\in[0,T_0]} \|a(T)\|^4 + 1 \right) ,$$

which is bounded, as $\mathbb{E}\sup_{T\in[0,T_0]} \|a(T)\|^{4p} \le C$ is bounded by standard a priori estimates for a (See e.g. Lemma B.9).

Finally, by the maximal inequality of Lemma A.4 for $p > 4$

$$\mathbb{E} \sup_{T\in[0,T_0]} \left\| \int_0^T e^{(T-\tau)L\varepsilon^{-2}} G_2(\tau)d\tilde{\beta}(\tau) \right\|^p \le C\varepsilon^{-2}\varepsilon^p \cdot \sup_{T\in[0,T_0]} \mathbb{E}\|G_2(T)\|^p \, .$$

Using again the bound $a = \mathcal{O}(1)$ from Lemma B.9, the claim follows. $\qquad \square$

Proof. (of Theorem 2.6) Suppose u is a mild solution (i.e. a solution of (2.51)) such that all moments are finite. To be more precise suppose that for all $p \ge 1$ there is a constant C_p such that

$$\mathbb{E}\|a(0)\|^p \le C_p \quad \text{and} \quad \mathbb{E}\|\psi(0)\|^p \le C_p \, ,$$

where a and ψ are defined in (2.59). We discuss the general case later at the end of the proof.

Define $R_c \in \mathcal{N}$ and $R_s \in P_s X$ by

$$u(t) = \varepsilon w(t) + \varepsilon^2 R_c(t) + \varepsilon^3 R_s(t) \, .$$

For short-hand notation define $R = R_c + \varepsilon R_s$. It is obvious, that by definition $R_c(0) = 0$ and $R_s(0) = 0$. Using the mild formulation (2.51) and the residual (2.52) we derive

$$\varepsilon^2 R(t) = \int_0^t e^{(t-\tau)L} \varepsilon^4 AR(\tau)d\tau + \int_0^t e^{(t-\tau)L} \varepsilon^3 R(\tau)d\beta(\tau)$$

$$+ \int_0^t e^{(t-\tau)L}(B(u) - \varepsilon^2 B(w))(\tau)d\tau + \mathrm{Res}(\varepsilon w)(t) \ .$$

Expanding $B_c(u) - \varepsilon^2 B_c(w)$ and using $B_c \equiv 0$ on $\mathcal{N} \times \mathcal{N}$, yields for R_c

$$R_c(t) = \int_0^t \varepsilon^2 A_c R_c(\tau) + \varepsilon^3 A_c R_s(\tau)d\tau + \varepsilon^{-2} P_c \mathrm{Res}(\varepsilon w)(t)$$

$$+ 2\varepsilon^2 \int_0^t (B_c(\psi, R_c) + B_c(a, R_s))(\tau)d\tau + \int_0^t \varepsilon R_c(\tau)d\beta(\tau)$$

$$+ \varepsilon^3 \int_0^t (2B_c(\psi, R_s) + 2B_c(R_c, R_s) + \varepsilon B_c(R_s))(\tau)d\tau \ . \qquad (2.70)$$

For R_s we obtain

$$R_s(t) = \int_0^t e^{(t-\tau)L} \varepsilon A_s(R_c + \varepsilon R_s)(\tau)d\tau + \varepsilon^{-3} P_s \mathrm{Res}(\varepsilon w)(t)$$

$$+ \int_0^t e^{(t-\tau)L}(2B_s(a, R_c) + 2\varepsilon B_s(\psi, R_c) + 2\varepsilon B_s(w, R_s) + \varepsilon B_s(R))(\tau)d\tau$$

$$+ \int_0^t e^{(t-\tau)L} \varepsilon R_s(\tau)d\beta(\tau) \ . \qquad (2.71)$$

In this system of equations (2.70) and (2.71), we introduce a cut-off. Take ρ as in the proof of Theorem 2.4, and define the cut-off for the operators by

$$A^{(\rho)}(z) = \rho(\|z\|\varepsilon^{1/2})Az \quad \text{and} \quad B^{(\rho)}(z_1, z_2) = \rho(\|z_1\|\varepsilon^{1/2})\rho(\|z_2\|\varepsilon^{1/2})B(z_1, z_2)$$

for all z, $z_i \in X$. Additionally, we use a cut-off for the diffusion term in (2.71)

$$\sigma^{(\rho)}(z) = \rho(\|z\|\varepsilon^{1/2})z \ ,$$

in order to replace the R_s in the stochastic integral in (2.71).

Finally replace a and ψ by $a^{(\rho)}$ and $\psi^{(\rho)}$, where for $1 \ll \gamma_\varepsilon \ll \varepsilon^{-1/2}$ to be fixed later

$$a^{(\rho)} := \begin{cases} a : \text{for } \|a\| \leq \gamma_\varepsilon \\ 0 : \text{otherwise} \ . \end{cases}$$

Let $\psi^{(\rho)}$ as in (2.46), but now defined with $a^{(\rho)}$ and

$$\psi^{(\rho)}(0) := \begin{cases} \psi(0) : \text{for } \|\psi(0)\| \leq \gamma_\varepsilon^2 \\ 0 \ : \text{otherwise} \ . \end{cases}$$

It is easy to verify that $\|\psi^{(\rho)}(t)\| \leq C\gamma_\varepsilon^2$ for all $t \in [0, T_0\varepsilon^{-2}]$.

Now let $(R_c^{(\rho)}, R_s^{(\rho)})$ be the solution of the modified equations (2.70) and (2.71) with cut-off $A^{(\rho)}$, $B^{(\rho)}$, $\sigma^{(\rho)}$, $a^{(\rho)}$, and $\psi^{(\rho)}$. Note that for simplicity, we do not change the residual. Now

$$\|R_s^{(\rho)}(t)\| \le C + \varepsilon^{-3}\|P_s\mathrm{Res}(\varepsilon w)(t)\| + C\gamma_\varepsilon \int_0^t \|R_c^{(\rho)}(\tau)\| e^{-(t-\tau)\omega}(t-\tau)^{-\alpha} d\tau$$
$$+\varepsilon\left\| \int_0^t e^{(t-\tau)L}\sigma^{(\rho)}(R_s^{(\rho)}(\tau))d\beta(\tau) \right\|. \tag{2.72}$$

We can choose q sufficiently large and $t \le T_0\varepsilon^{-2}$ such that by Theorem 2.5 (we need $\mathbb{E}\|a(0)\|^{4q} < \infty$ and $\mathbb{E}\|\psi(0)\|^{2q} < \infty$) together with Hölder, Theorem 2.5, and maximal inequality (cf. Lemma A.4) we can split

$$\|R_s^{(\rho)}(t)\| \le C\gamma_\varepsilon\|G_2(t)\| + \|G_1(t)\|, \tag{2.73}$$

where the G_i are such that for all small $\kappa > 0$.

$$\mathbb{E}\left(\sup_{t\in[0,T_0\varepsilon^{-2}]} \|G_1(t)\|^q \right) \le C\varepsilon^{-\kappa} \quad \text{and} \quad \|G_2(t)\| \le \sup_{\tau\in[0,t]} \|R_c^{(\rho)}(\tau)\|.$$

The term $\varepsilon^{-\kappa}$ arises by using Hölder's inequality from taking arbitrary high moments of the residual. For instance, we can choose $\kappa \in (0, \frac{1}{4}]$ arbitrary. Additionally,

$$\mathbb{E}\left(\|R_c^{(\rho)}(t)\|^{q-1}\|G_2(t)\| \right) \le \int_0^\infty \tau^{-\alpha}e^{-\tau\omega}d\tau \cdot \sup_{\tau\in[0,t]} \mathbb{E}\|R_c^{(\rho)}(\tau)\|^q.$$

From the modified version of (2.70) for $t \le T_0\varepsilon^{-2}$

$$dR_c^{(\rho)} = \varepsilon^2\left(A_c^{(\rho)}R_c^{(\rho)} + 2B_c^{(\rho)}(\psi^{(\rho)}, R_c^{(\rho)}) + 2B_c^{(\rho)}(a^{(\rho)}, R_s^{(\rho)}) + \mathcal{G} \right)dt$$
$$+\varepsilon R_c^{(\rho)}d\beta + \varepsilon^2(A_c\psi(\varepsilon^2 t) + B_c(\psi(\varepsilon^2 t)))dt. \tag{2.74}$$

where we used that $P_c\mathrm{Res}$ is differentiable by Theorem 2.5. The term $\mathcal{G}$ collects terms of order $\mathcal{O}(1)$. To be more precise

$$\mathbb{E} \sup_{t\in[0,T_0\varepsilon^{-2}]} \|\mathcal{G}(t)\|^q \le C.$$

Denote by $\tilde{R}_c^{(\rho)}(T) = R_c^{(\rho)}(t)$ and $\tilde{R}_s^{(\rho)}(T) = R_s^{(\rho)}(t)$, respectively, the corresponding processes on the slow time-scale $T = \varepsilon^2 t$. Then we can rescale (2.74) to derive

$$d\tilde{R}_c^{(\rho)} = \left(A_c^{(\rho)}\tilde{R}_c^{(\rho)} + 2B_c^{(\rho)}(\psi^{(\rho)}, \tilde{R}_c^{(\rho)}) + 2B_c^{(\rho)}(a^{(\rho)}, \tilde{R}_s^{(\rho)}) + \tilde{\mathcal{G}} \right)dT$$
$$+\tilde{R}_c^{(\rho)}d\tilde{\beta} + (A_c\psi + B_c(\psi))\,dT, \tag{2.75}$$

where $\tilde{\beta}$ is as usual the rescaled Brownian motion. Now we can consider $\tilde{R}_s^{(\rho)}$ as a function of $\tilde{R}_c^{(\rho)}$ with bounds derived in (2.73). Define

$$f(T, \tilde{R}_c^{(\rho)}) = A_c^{(\rho)}\tilde{R}_c^{(\rho)} + 2B_c^{(\rho)}(\psi^{(\rho)}, \tilde{R}_c^{(\rho)}) + 2B_c^{(\rho)}(a^{(\rho)}, \tilde{R}_s^{(\rho)})$$

and

$$g(T) = \tilde{\mathcal{G}} + A_c \psi + B_c(\psi) \, .$$

Furthermore, using (2.73) and a direct estimate neglecting the cut-off

$$\|f(T, \tilde{R}_c^{(\rho)}(T))\| \leq C\gamma_\varepsilon \|R_s^{(\rho)}(T)\| + C\gamma_\varepsilon^2 \|R_c^{(\rho)}(T)\|$$
$$\leq C\gamma_\varepsilon^2 \|\tilde{G}_2(T)\| + C\gamma_\varepsilon \|G_1(T)\| + C\gamma_\varepsilon^2 \|R_c^{(\rho)}(T)\|$$

and for $q \geq 2$

$$\mathbb{E}\|\tilde{R}_c^{(\rho)}(T)\|^{q-1}\|f(T, \tilde{R}_c^{(\rho)}(T))\|$$
$$\leq C\gamma_\varepsilon^2 \sup_{\tau \in [0,T]} \mathbb{E}\|\tilde{R}_c^{(\rho)}(\tau)\|^q + C\gamma_\varepsilon \mathbb{E}\|\tilde{G}_1(T)\|\|\tilde{R}_c^{(\rho)}(T)\|^{q-1}$$
$$\leq C\gamma_\varepsilon^2 \sup_{\tau \in [0,T]} \mathbb{E}\|\tilde{R}_c^{(\rho)}(\tau)\|^q + C\varepsilon^{-\kappa} \, .$$

First using the usual bounds on a from Lemma B.9 implies (needs $\mathbb{E}\|a(0)\|^{2q} < C$ and $\mathbb{E}\|\psi(0)\|^q < C$)

$$\mathbb{E}\left(\sup_{T \in [0,T_0]} \|g(T)\|^q \right) \leq C \, .$$

To proceed consider $\tilde{R}_c^{(\rho)}$ as a solution of $dx = (f(\cdot, x) + g)dT + x d\tilde{\beta}$. We already verified all assumptions of Lemma B.10. This lemma provides a priori bounds for SDEs, which are exactly of the type needed for $R_c^{(\rho)}$. It yields

$$\mathbb{E} \sup_{T \in [0,T_0\varepsilon^{-2}]} \|R_c^{(\rho)}(t)\|^q \leq C\gamma_\varepsilon^{q+2} e^{c_q \gamma_\varepsilon^2} \varepsilon^{-\kappa}$$
$$\leq C\varepsilon^{-2\kappa} \, , \tag{2.76}$$

where we choose $\gamma_\varepsilon^2 = \gamma \ln(\varepsilon^{-1})$ for some sufficiently small $\gamma = \gamma(q) > 0$. Note that the constant c_q depends explicitly on q. It is essential that $c_q \gamma \leq \kappa$.

Furthermore, using (2.73), yields for $R_s^{(\rho)}$

$$\mathbb{E}\left(\sup_{t \in [0,T_0\varepsilon^{-2}]} \|R_s^{(\rho)}(t)\|^q \right) \leq C\varepsilon^{-3\kappa} \, . \tag{2.77}$$

Define the stopping time τ_R^* as the time such that (R_c, R_s) and $(R_c^{(\rho)}, R_s^{(\rho)})$ coincide up to τ_R^*. Thus $R_s^{(\rho)}(t) = R_s(t)$ and $R_c^{(\rho)}(t) = R_c(t)$ for all $t \leq \tau_R^*$, and

$$\mathbb{P}\left(\sup_{t \in [0, \frac{T_0}{\varepsilon^2}]} \|u(t) - \varepsilon w(t)\| \leq C\varepsilon^{2-4\kappa} \right)$$

$$\geq \mathbb{P}\left(\sup_{t \in [0, \frac{T_0}{\varepsilon^2}]} \|R_c(t)\| \leq C\varepsilon^{-4\kappa}, \quad \sup_{t \in [0, \frac{T_0}{\varepsilon^2}]} \|R_s(t)\| \leq C\varepsilon^{-1-4\kappa} \right)$$

$$\geq \mathbb{P}\left(\tau_R^* \leq T_0 \varepsilon^{-2}, \quad \sup_{t \in [0, \frac{T_0}{\varepsilon^2}]} \|R_c^{(\rho)}(t)\| \leq C\varepsilon^{-4\kappa}, \quad \sup_{t \in [0, \frac{T_0}{\varepsilon^2}]} \|R_s^{(\rho)}(t)\| \leq C\varepsilon^{-1-4\kappa} \right)$$

$$\geq \mathbb{P}\left(\tau_R^* \leq T_0 \varepsilon^{-2} \right) - C\varepsilon^{2\kappa(2q-1)} - C\varepsilon^{q+2\kappa(4q-3)}, \tag{2.78}$$

where we used Chebychev's inequality, (2.76), and (2.77). Furthermore, we can fix any $\delta_1 > 0$ such that

$$\mathbb{P}\left(\tau_R^* \leq T_0 \varepsilon^{-2} \right)$$

$$\geq \mathbb{P}\left(\sup_{T \in [0, T_0]} \|a(T)\| \leq \gamma_\varepsilon, \quad \sup_{t \in [0, \frac{T_0}{\varepsilon^2}]} \|R_c^{(\rho)}(t)\| \leq C\varepsilon^{-\frac{1}{2}}, \right.$$

$$\left. \sup_{t \in [0, \frac{T_0}{\varepsilon^2}]} \|R_s^{(\rho)}(t)\| \leq C\varepsilon^{-\frac{1}{2}}, \quad \|\psi(0)\| \leq \gamma_\varepsilon^2 \right)$$

$$\geq 1 - \mathbb{P}\left(\|\psi(0)\| > \gamma_\varepsilon^2 \right) - \mathbb{P}\left(\|a(0)\| > \delta_1 \right) - C\varepsilon^\eta \tag{2.79}$$

for some small $\eta > 0$, where we used Chebychev, a large deviation estimate for a (see e.g. Theorem B.9), and again (2.76) and (2.77). Recall that we fixed $\gamma_\varepsilon^2 = \gamma \ln(\varepsilon^{-1})$ for some small $\gamma > 0$.

Hence, provided $\mathbb{E}\|a(0)\|^{\tilde{q}} \leq C$ and $\mathbb{E}\|\psi(0)\|^{\tilde{q}} \leq C$ for some sufficiently large $\tilde{q} > 0$, we proved that given small $\kappa > 0$ and positive constants δ_1, δ_2 there is a constant $C > 0$ and some small $\eta > 0$ such that

$$\mathbb{P}\left(\sup_{t \in [0, \frac{T_0}{\varepsilon^2}]} \|u(t) - \varepsilon w(t)\| \leq C\varepsilon^{2-4\kappa} \right)$$

$$\geq 1 - \mathbb{P}\left(\|u_s(0)\| > \delta_2 \varepsilon^2 \right) - \mathbb{P}\left(\|u_c(0)\| > \delta_1 \varepsilon \right) - C\varepsilon^\eta .$$

If we take another cut-off for the initial condition $u(0)$, then it is easy to finish the proof of Theorem 2.6. $\qquad \square$

2.5 Setting for Additive Noise (Thermal Noise)

In this section, we follow partly the presentation in [BH05], which reviews results of [Blö03] and [BH04], which in turn are based on [BMPS01]. The setting is exactly the one sketched in Section 1.2. We focus on an SPDE of the type (1.2) with mild

solutions given by (1.19). The additive noise $\varepsilon^2\xi$ in the equation is for instance
motivated by the presence of thermal fluctuations in the medium. Therefore the
strength ε^2 of the noise is supposed to be very small. We usually assume that the
noise $\xi = \partial_t W$ is some generalised Gaussian process, which is given by the derivative
of some Q-Wiener process. We comment on that in detail later after Assumption
2.8.

In Section 2.5.1 we summarise the precise mathematical assumptions for (1.2).
The main results for the approximation via amplitude equations are given in Section
2.5.3. This setting is also used in Section 3.1 to approximate long-time behaviour
in terms of invariant measures.

2.5.1 *Assumptions*

Let us summarise all assumptions necessary for our results. We do not focus on the
highest possible level of generality, but stick to some simpler setting which cover all
our examples. First consider the linear operator L.

Assumption 2.5   *Let X be a separable Hilbert space and Δ (subject to some
boundary conditions on a bounded domain) be a self-adjoint version of the Laplacian
on X. Suppose $L = P(-\Delta)$ for some function P such that L is non-positive.
Furthermore, let the kernel $\mathcal{N} = \ker L$ of L be non-empty and finite dimensional.
Finally, suppose $P(k) \to -\infty$ as $k \to \infty$.*

This assumption is a stronger than the one in Section 2.2. It is mainly used for
convenience of presentation, and covers all examples presented. Furthermore, it
is just a special case of Assumption 2.1, and in the following we can use all the
implications of this assumption. We use the notation P_c and P_s, which are in this
case just the standard orthogonal projections. Additionally, recall the splitting
$X = \mathcal{N} \oplus \mathcal{S}$ with $\mathcal{S} = P_s X$ and the spaces X^α from Section 2.2. Recall furthermore
the bounds (2.4), (2.5), and (2.6) for the analytic semigroup e^{tL} generated by L.

For the nonlinearities, we make two assumptions. The first one, is much weaker
than Assumption 2.2, as we are aiming only for local results in that case. Especially,
we can get rid of the strong nonlinear dissipativity. The second assumption is
similar to Assumption 2.2 and involves strong nonlinear stability and dissipativity
conditions in $\mathcal{N}$.

Assumption 2.6   *The function $\mathcal{F}$ is locally Lipschitz from X to $X^{-\alpha}$ for some
$\alpha \in [0, 1)$. This means that for all $R > 0$ there is a $C > 0$ such that*

$$\|\mathcal{F}(v_1) - \mathcal{F}(v_2)\|_{-\alpha} \le C\|v_1 - v_2\| \quad \text{for all } v_i \text{ with } \|v_i\| \le R.$$

*Assume we can split $\mathcal{F}(x) = f(x) + g(x)$, where $f : X \times X \times X \to X^{-\alpha}$ is con-
tinuous, trilinear, and symmetric. The function g is of higher order, which means
$\|g(x)\|_{-\alpha} \le C\|x\|^4$ provided $\|x\| \le 1$.*

Furthermore, assume that $P_c f$ is stable on $\mathcal{N}$, which means that for some small $\delta > 0$ there is a constant $C_\delta > 0$ such that $|\langle P_c f(\varphi + a), a\rangle| \leq C_\delta \|\varphi\|^4 - \delta\|a\|^4$ holds for all $a, \varphi \in \mathcal{N}$, where we use the canonical scalar product in $\mathcal{N} \approx \mathbb{R}^n$. Finally, we assume that $A : X \to X^{-\alpha}$ is a bounded linear operator.

As usual, we use the shorthand notations $f(u) = f(u, u, u)$ and $\mathcal{F}_c = P_c \mathcal{F}$, Moreover $A_c = P_c A$ and $f_c = P_c f$.

In the following we make a somewhat stronger assumption ensuring global non-linear stability of our SPDE (1.2). It is similar to Assumption 2.2. Nevertheless, we state it here, as it is slightly different. For simplicity, we restrict ourselves to cubic nonlinearities. This assumption is responsible for the global existence of solutions (cf. Proposition 2.1) and for uniform (in t) bounds on $\mathbb{E}\|u(t)\|^p$ for solutions of (1.2) (cf. Theorem 2.8). As before these bounds are independent of the initial condition.

Assumption 2.7   *Let Assumption 2.6 be true and assume that the linear operator A belongs to $\mathcal{L}(X^1, X)$. Moreover, there exists a constant $C_A > 0$ such that*

$$\langle Av, v\rangle \leq C_A\big(\|v\|^2 + \langle -Lv, v\rangle\big) \quad \text{for all} \quad v \in X^1 \,. \tag{2.80}$$

We also assume that $\mathcal{F}$ is trilinear, $\mathcal{F} : (X^1)^3 \to X$ is continuous and that

$$\langle \mathcal{F}_c(v_c, v_c, w_c), w_c\rangle < 0 \quad \text{for all } v_c, w_c \in \mathcal{N} \setminus \{0\} \,. \tag{2.81}$$

We finally assume that there exist constants K, $\delta > 0$, and $\gamma_L \in [0, 1)$ such that,

$$\langle \mathcal{F}(v + \phi), v\rangle \leq K\|\phi\|^4 - \delta\|v\|^4 - \gamma_L\langle Lv, v\rangle \tag{2.82}$$

for all $\phi, v \in X^1$.

Concerning the stochastic perturbation we always assume that the following is true. For a detailed discussion of Q-Wiener processes and stochastic convolutions see [DPZ92] and the remark below.

Assumption 2.8   *The noise process is formally given by $\xi = Q\partial_t W$, where W is a standard cylindrical Wiener process in X with the identity as a covariance operator and $Q \in \mathcal{L}(X, X)$ is symmetric. Furthermore, there exists a constant $\tilde{\alpha} < \frac{1}{2}$ such that*

$$\|(1 - L)^{-\tilde{\alpha}} Q\|_{\mathrm{HS}(X)} < \infty \,, \tag{2.83}$$

where $\|\cdot\|_{\mathrm{HS}(X)}$ denotes the Hilbert-Schmidt norm of an operator from X to X.

Remark 2.7   *Straightforward computations, combined with the properties of analytic semigroups allow to check that Assumption 2.8 implies the following (see [DPZ92, Section 5.4] for the first assertion):*

- *The stochastic convolution $W_L(t) = \int_0^t e^{L(t-s)} Q\, dW(s)$ is an X-valued process with Hölder continuous sample paths.*

- *There exist positive constants C and γ such that*

$$\|P_s e^{Lt} Q\|_{\mathrm{HS}(X)} \leq C(1 + t^{-\gamma}) e^{-\omega t} \tag{2.84}$$

  holds for every $t > 0$.

Let us comment in more detail on the relationship between the Wiener process QW and the noise ξ. The standard cylindrical Wiener-process W is given as $W(t) = \sum_{k=1}^{\infty} \beta_k(t) e_k$, where $\{\beta_k\}_{k \in \mathbb{N}}$ is any family of independent identically distributed standard Brownian motions, and $\{e_k\}_{k \in \mathbb{N}}$ is any orthonormal basis of X. Obviously, $W(t)$ is not defined in X, but in some larger space.

Let ξ be a generalised Gaussian process such that

$$\mathbb{E}\xi(t, x) = 0 \quad \text{and} \quad \mathbb{E}\xi(t, x)\xi(s, y) = \delta(t - s)q(x, y) \,,$$

where δ is the usual Delta-distribution and q the spatial correlation-function.

Assume for example that $X = L^2(G)$. If we define the linear operator Q via the convolution $Q^2 f(x) = \int_G q(x, y) f(y) dy$, then it is easy to verify that the generalised derivative $\partial_t QW$ has the same properties as ξ. A technical point is that we need to extend Q to the space, where W is defined. See for instance [Blö05b] and the references therein.

Looking at (2.86) in Proposition 2.1, we see that the noise acting on $\mathcal{N}$ and $\mathcal{S}$ could be highly correlated (cf. Remark 2.8). Nevertheless it acts completely different in $\mathcal{N}$ and $\mathcal{S}$. It is obvious that

$$P_s[W_L(t)] = \int_0^t e^{(t-\tau)L} dP_s QW(\tau) \quad \text{and} \quad P_c[W_L(t)] = P_c QW(t) \,. \tag{2.85}$$

Therefore, the stochastic convolution is a Wiener process on $\mathcal{N}$ and it is a stable Ornstein-Uhlenbeck process on $\mathcal{S}$. As already discussed in (1.20) and (1.21), the noise acts in two completely different ways on $P_c u$ and $P_s u$ for some mild solution u.

Remark 2.8 *Note that we do not assume that Q and L commute. Hence, it is in general not true that Q and P_c commute. Therefore, the Wiener processes $P_c QW$ and $P_s QW$ are not necessarily independent, which implies that the amplitude equation (1.5) and equation (1.6) for the second order correction are coupled through the noise. Nevertheless it is a key point that the two noise terms in both equations live on different time-scales.*

2.5.2 *Existence of Solutions*

To give a meaning to (1.2) we always consider mild solutions, which are given by the following proposition (cf. [BH04]).

Proposition 2.1 *Under Assumption 2.5, 2.6, and 2.8, for all (stochastic) initial conditions $u(0) \in X$ equation (1.2) has a unique local mild solution u. This means* *we have a stopping time $t^* > 0$ and a stochastic process u such that $u : [0, t^*) \to X$ is continuous and fulfils*

$$u(t) = e^{tL}u(0) + \int_0^t e^{(t-\tau)L}[\varepsilon^2 Au + \mathcal{F}(u)](\tau)d\tau + \varepsilon^2 W_L(t) \qquad (2.86)$$

for $0 < t < t^*$. If we suppose additionally that Assumption 2.7 holds, then the solutions are global, which means $t^* = \infty$.

We will not give a detailed proof here, as the existence and uniqueness of local solutions is standard since we consider locally Lipschitz-continuous nonlinearities. See for example [DPZ92, Section 7]. Using the regularity of the stochastic convolution we can also apply the deterministic approach of [Hen81, Thm. 3.3.3] path-wise.

The global existence follows from standard a priori estimates for $v = u - \varepsilon^2 W_L$, as v is a weak solution of the following PDE with random coefficients

$$\partial_t v = Lv + \varepsilon^2 A(v + \varepsilon^2 W_L) + \mathcal{F}(v + \varepsilon^2 W_L). \qquad (2.87)$$

The formal idea is to take the scalar product with v, in order to derive standard a priori estimates for $\|v\|^2$ and hence $\|u\|^2$. This is for instance the reason, why we assume that $A \in \mathcal{L}(X^1, X)$ and $\mathcal{F} \in \mathcal{L}_3(X^1, X)$ in Assumption 2.7, which was not present in Assumption 2.2. This is essential for getting higher regularity, which is necessary in order to make the a priori estimates rigorous.

For multiplicative noise, we avoided this by assuming that the solution is strong. Thus we could directly apply Itô's formula. See for example the proof of Theorem 2.4. For additive noise we can in general not apply Itô's formula. This is only possible for trace-class noise. Neither can we use a priori estimates directly for the random PDE (2.87). We need an argument relying on sufficiently smooth approximations v.. See for instance the proof of Theorem 4.1 of [BH04].

2.5.3 *Amplitude Equations — Main Results*

We review two approaches to verify the approximation via amplitude equations. One relies on a purely local picture and uses Assumption 2.6. This is similar to the results of [BMPS01] or [Blö03]. The second approach was developed in [BH04]. It takes into account the global nonlinear stability of the equation given by Assumption 2.7. This latter approach is also similar to the results of Section 2.2. One main difference is that we cannot use Itô's formula. We rely on a priori estimates for (2.87) and bounds on the stochastic convolution W_L. For the presentation we follow partly the review article [BH05].

2.5.3.1 *Attractivity*

The attractivity justifies the ansatz for the formal computation. It shows that after a comparably short time the solution is of the form of the ansatz (1.4). We first state the local result relying on solutions of order $\mathcal{O}(\varepsilon)$.

Theorem 2.7 (Attractivity-local) *Under Assumptions 2.5, 2.6, and 2.8 fix some small constant $\kappa > 0$. Then there are constants $c_i > 0$ and a time $t_\varepsilon = \mathcal{O}(\ln(\varepsilon^{-1}))$ such that for all mild solutions u of (2.86) we can write*

$$u(t_\varepsilon) = \varepsilon a_\varepsilon + \varepsilon^2 R_\varepsilon \quad \text{with} \quad a_\varepsilon \in \mathcal{N} \text{ and } R_\varepsilon \in \mathcal{S},$$

where

$$\mathbb{P}\Big(\|a_\varepsilon\| \leq \delta, \ \|R_\varepsilon\| \leq \varepsilon^{-\kappa}\Big) \geq \mathbb{P}\Big(\|u(0)\| \leq c_3 \delta \varepsilon\Big) - c_1 \mathrm{e}^{-c_2 \varepsilon^{-2\kappa}} \tag{2.88}$$

for all $\delta > 1$ and $\varepsilon \in (0,1)$.

This result states in a weak form that $u(0) = \mathcal{O}(\varepsilon)$ with high probability implies $P_c u(t_\varepsilon) = \mathcal{O}(\varepsilon)$ and $P_s u(t_\varepsilon) = \mathcal{O}(\varepsilon^2)$ with high probability, too. Note that we do not bound any moments of the solution u.

We do not give a detailed proof of this result, as it is a straightforward modification of Theorem 3.3 of [Blö03]. It relies on the fact that small solutions of order $\mathcal{O}(\varepsilon)$ are on small time-scales given by the linearised picture, which is dominated by the semigroup estimates (2.5) and (2.6). Thus modes in $P_s X$ decay exponentially fast on a time-scale of order $\mathcal{O}(1)$.

Using strong nonlinear stability, we can prove much more:

Theorem 2.8 (Attractivity-global) *Let Assumptions 2.5, 2.6, and 2.8 be satisfied. Then for all times $T_\varepsilon = T_0 \varepsilon^{-2} > 0$ and for all $p \geq 1$ there are constants $C_p > 0$ explicitly depending on p such that*

$$\mathbb{E}\|u(t + T_\varepsilon)\|^p \leq C_p \varepsilon^p \quad \text{and} \quad \mathbb{E}\|P_s u(t + T_\varepsilon)\|^p \leq C_p \varepsilon^{2p} \tag{2.89}$$

for all $t \geq 0$, all X-valued mild solutions u of equation (1.2) independent of the initial condition $u(0)$, and for all $\varepsilon \in (0,1)$.

Furthermore, if we already assume that $\mathbb{E}\|u(0)\|^p \leq \tilde{C}_p \varepsilon^p$ for a constant $\tilde{C}_p > 0$, then there is a time $t_\varepsilon = \mathcal{O}(\ln(\varepsilon^{-1}))$ and a constant $C > 0$ such that

$$\mathbb{E}\|u(t)\|^p \leq C \varepsilon^p \quad \text{and} \quad \mathbb{E}\|P_s u(t + t_\varepsilon)\|^p \leq C \varepsilon^{2p} \tag{2.90}$$

for all $t \geq 0$, all X-valued mild solutions u, and for all $\varepsilon \in (0,1)$.

The proof is given by a priori estimates. This was not directly proved in [BH04], but under our somewhat stronger assumptions this is similar to Lemma 4.3 of [BH04]. It relies on a priori estimates for $v_{\delta_\varepsilon} = u - \varepsilon^2 W_{L-\delta_\varepsilon}$ with $\delta_\varepsilon = \mathcal{O}(\varepsilon^2)$, which fulfils a random PDE similar to (2.87). The main technical advantage is that the linear semigroup generated by $L - \delta_\varepsilon$ is exponentially stable.

2.5.3.2 *Approximation*

For a solution a of (1.5) and ψ of (1.6) we define the approximations εw_k of order k by

$$\varepsilon w_1(t) := \varepsilon a(\varepsilon^2 t) \quad \text{and} \quad \varepsilon w_2(t) := \varepsilon a(\varepsilon^2 t) + \varepsilon^2 \psi(t).$$

The residual of εw is given by (1.23). Now the main idea is to obtain bounds on $P_c\mathrm{Res}(\varepsilon w)$ via the amplitude equation and to bound $P_s\mathrm{Res}(\varepsilon w)$ by using the stability of the equation which is ensured by our spectral gap (cf. (2.4) or (2.5)). As usual, these estimates require good a priori bounds on the approximation εw_k, but do not require any a priori knowledge on the solution u of the original equation.

Bounds on the residual easily imply approximation results, as we can establish bounds on the difference of εw_k and u using (1.23) and (2.86).

Theorem 2.9 **(Approximation-local)** *Suppose Assumptions 2.5, 2.6, and 2.8 are true. Fix the time $T_0 > 0$ and some small $\kappa \in (0,1)$. Then there are constants $C_{\mathrm{att}} > 0$ and $c_i > 0$ such that for $\varepsilon \in (0,1)$ we obtain for all solutions u of (2.86) and all solutions a of (1.5) (with f_c instead of $\mathcal{F}_c$)*

$$\mathbb{P}\left(\sup_{t\in[0,T_0\varepsilon^{-2}]} \|u(t) - \varepsilon w_1(t)\| \leq C_{\mathrm{att}}\varepsilon^{2-\kappa} \right) \tag{2.91}$$

$$\geq 1 - \mathbb{P}\left(\|u(0) - \varepsilon a(0)\| \geq c_1\varepsilon^{2-\kappa} \right) - \mathbb{P}\left(\|u(0)\| \geq c_2\varepsilon \right) - c_3\mathrm{e}^{-c_4\ln(\varepsilon^{-1})^2} .$$

The proof of this result is a straightforward modification of Theorems 4.1 and 4.3 of [Blö03]. We use ideas of [BHP05] to allow for weaker bounds on $\sup_{T\in[0,T_0]}|a(T)|$ by $c^*\ln(\varepsilon^{-1})$, which were not present in [Blö03] or [BMPS01]. There the probability was bounded by terms that are small for $\varepsilon \to 0$, without any further information on the order of the smallness. Nevertheless, we can easily improve these results. In that situation, it is enough to use large deviation bounds for $\sup_{T\in[0,T_0]}|a(T)|$, in order to bound this term by $\gamma\ln(\varepsilon^{-1})$ with high probability. This is similar to Theorem B.9, but now we need to carry over large deviation results for the driving Brownian motion to the solution of the SDE.

As before for multiplicative noise, if we assume nonlinear stability, as for instance given by Assumption 2.7, then we can prove a much stronger result, where we can bound all moments.

Theorem 2.10 **(Approximation-global)** *Suppose Assumptions 2.5, 2.7, and 2.8 hold and let u be the mild solution of (1.2) with (random) initial value $u(0)$, which fulfils (2.89). This means there is a family of positive constants $\{C_p\}_{p\geq 1}$ such that*

$$\mathbb{E}\|u(0)\|^p \leq C_p\varepsilon^p \quad and \quad \mathbb{E}\|P_s u(0)\|^p \leq C_p\varepsilon^{2p} . \tag{2.92}$$

Then for all $p \geq 1$, $1 \gg \kappa > 0$ and $T_0 > 0$ there is a constant $C_{\mathrm{app}} > 0$ explicitly depending on p and T_0 such that the estimate

$$\mathbb{E}\left(\sup_{t\in[0,T_0\varepsilon^{-2}]} \|u(t) - \varepsilon w_2(t)\|^p \right) \leq C_{\mathrm{app}}\varepsilon^{3p-\kappa}$$

holds for $\varepsilon \in (0,1)$.

The proof is Corollary 3.9 of [BH04].

2.6   Quadratic Nonlinearities

In this section we review the results of [Blö05a]. Consider an SPDE of the following type.

$$\partial_t u = Lu + \varepsilon^2 Au + B(u, u) + \varepsilon^2 \xi \,, \qquad (2.93)$$

where L as in Assumption 2.1. The linear operator A and the bilinear operator B are as in Assumption 2.4, and the noise is the generalised derivative of some Wiener process (cf. Assumption 2.9).

In [Blö05a] we used fractional noise. This was motivated by the fact that the proofs rely on fractional integration by parts formulas, and explicit path-wise estimates. Here we state for simplicity only the version for Gaussian noise that is white in time. Note that due to the method of proof, we need the noise to be trace-class, as we need bounds for the Wiener process $W(t)$ in the space X. This obviously rules out space-time white noise.

Let us furthermore point out that the Hilbert space setting is not necessary in this approach, as we purely rely on local results, using cut-off techniques, and we do not use a priori estimates. It is also necessary to deal with non self-adjoint operators, as the linear part in the Rayleigh-Bénard system is not self-adjoint (cf. Section 6.1 of [Blö05a] for a detailed discussion.

For the stochastic perturbation ξ let the following assumption be true.

Assumption 2.9   **(Noise)** *Suppose that the noise process ξ is the generalised derivative of some Wiener process $\{QW(t)\}_{t\geq 0}$ on some probability space $(\Omega, \mathcal{F}, \mathbb{P})$, where W is the standard cylindrical Wiener process.*

Assume that the stochastic convolution

$$W_L(t) = \int_0^t e^{(t-\tau)L} dQW(\tau) \qquad (2.94)$$

is a well defined stochastic process with continuous paths in X. We suppose that the noise (or W) is of trace-class, i.e. $\operatorname{tr}(Q^2) = \mathbb{E}\|QW(t)\|^2 < \infty$.

This assumption is stronger than Assumption 2.8. Especially, W being trace-class is a serious restriction, as this already implies that W has continuous paths in X. We briefly sketched after Remark 2.7 the connection between the spatial correlation function q of the noise ξ and the operator Q belonging to W. The condition of W being trace-class is essentially a regularity condition on q. See for example [Blö05b]. Any decay condition for the eigenvalues of Q immediately transfers to a decay condition of the Fourier coefficients of q.

To give a meaning to (2.93) we consider as usual mild solutions.

Assumption 2.10   **(Mild Solutions)** *We assume that for all (stochastic) initial conditions $u(0) \in X$ equation (2.93) has a mild local solution u. This means we have a stopping time $t^* > 0$ and a stochastic process u such that $u : [0, t^*) \to X$ is* *continuous and u is* $\mathbb{P}$-*a.s. a solution of*

$$u(t) = e^{tL}u(0) + \int_0^t e^{(t-\tau)L}\left[\varepsilon^2 Au(\tau) + B(u(\tau))\right]d\tau + \varepsilon^2 W_L(t) \qquad \text{for } t \leq t^*. \quad (2.95)$$

Moreover, either $t^* = \infty$ *or* $\|u(t)\| \to \infty$ *for* $t \to t^*$.

This assumption is again mainly for convenience. Under Assumptions 2.1, 2.4, and 2.9 it is easy to verify it. See [DPZ92] for a textbook. For L^p-theory with application to the Navier-Stokes equation see for instance [BP99; BP00]. Moreover there are results on the Kuramoto-Sivashinsky equation in [DE01] and surface growth equation in [BG04]. As before, we can only expect local solutions to exist. In general there is no constant $T > 0$ such that $T \leq t^*$.

The amplitude equation was formally derived in (1.13). Here we state the rigorous formulation. Recall the *amplitude equation*

$$a(T) = a(0) + \int_0^T A_c a(s)ds - 2\int_0^T B_c\left(a(s), L_s^{-1}B_s(a(s))\right)ds + \beta(T), \quad (2.96)$$

where $\{\beta(T)\}_{T \geq 0}$ defined by $\beta(T) = \varepsilon P_c W(\varepsilon^{-2}T)$ is the usual Brownian motion in $\mathcal{N}$.

Our main results are a local attractivity result (see Theorem 2.11) and the approximation result (see Theorem 2.13).

Theorem 2.11 (**Attractivity**) *Suppose Assumptions 2.1, 2.4, and 2.9 are true. Fix the time* $t_\varepsilon = \frac{1}{\omega}\ln(\varepsilon^{-2})$ *with* ω *from (2.4). We can write all mild solutions of (2.93) with (random) initial conditions* $u(0)$ *as*

$$u(t_\varepsilon) = \varepsilon a_\varepsilon + \varepsilon^2 R_\varepsilon$$

with $a_\varepsilon \in \mathcal{N}$ *and* $R_\varepsilon \in P_s X$, *such that for all* $\delta > 0$ *there is a constant* $C = C(\delta) > 0$ *such that*

$$\mathbb{P}\left\{\|a_\varepsilon\| \leq C, \quad \|R_\varepsilon\| \leq \ln(\varepsilon^{-1})\right\} \geq \mathbb{P}\left\{\|u(0)\| \leq \delta\varepsilon\right\} - o_\varepsilon(1). \quad (2.97)$$

Here $o_\varepsilon(1)$ denotes a term that converges to 0 for $\varepsilon \to 0$. For simplicity, we will not focus on precise convergence rates. We already saw in a similar situation Section 2.4 that this is quite technical, as we need explicit bounds on various probabilities, and good knowledge how our constants depend on other constants.

Theorem 2.11 is proved as Theorem 2.13 in [Blö05a]. The result relies only the local linearised picture, without taking into account bounds for moments. It is very similar to Theorem 2.7. But here we do not have nonlinear stability at our disposal. Thus we cannot formulate a global result.

For a given solution u of (2.95) define the approximation εw by

$$\varepsilon w(t) := \varepsilon a(\varepsilon^2 t) + \varepsilon^2 \psi(t), \quad (2.98)$$

where a is a solution of the amplitude equation (2.96) with initial condition $a(0) = \varepsilon^{-1}P_c u(0)$. As in Section 2.4 we use a second order correction on the fast modes. Let $\psi(t) \in P_s X$ satisfy $\psi(0) = \varepsilon^{-2} P_s u(0)$ and

$$\psi(t) = e^{tL}\psi(0) + P_s W_L(t) + \int_0^t e^{(t-\tau)L} B_s(a(\varepsilon^2 \tau)) d\tau . \tag{2.99}$$

This is a slightly different idea than (2.46), which was used in Section 2.4. It is similar to (2.50), but here we have to take into account additionally stochastic effects due to the additive noise. But again, as in Section 2.4 we can simplify (2.99), by rescaling to the slow time-scale and identifying only terms of lowest order, using a formal Delta-distribution argument. See also Section 1.1.3. Nevertheless, for both multiplicative and additive noise we formally end up with

$$\psi(\varepsilon^{-2}T) \approx -L_s^{-1} B_s(a(T))$$

on the slow time-scale.

The residual of εw in this context is

$$\mathrm{Res}(\varepsilon w)(t) = -\varepsilon w(t) + \varepsilon^2 W_L(t)$$
$$+ e^{tL}\varepsilon w(0) + \int_0^t e^{(t-\tau)L}[\varepsilon^2 A \varepsilon w(\tau) + B(\varepsilon w(\tau))] d\tau \quad (2.100)$$

In order to show that εw is a good approximation of a solution u of (2.95), we have to control the residual. This is done in two steps. First $P_c \mathrm{Res}(\varepsilon w)$ is discussed, where we rely on the amplitude equation (2.96). The second step bounds $P_s \mathrm{Res}(\varepsilon w)$. The main tool is the equation for the second order correction (2.99). It is crucial for our result, that the latter bound is much better than the first one. We do not give a proof. See Theorem 2.14 of [Blö05a].

Theorem 2.12   (Residual) *Suppose Assumptions 2.1, 2.4, and 2.9 are true. Fix $\delta > 0$, some small $1 \gg \kappa \geq 0$, and some time $T_0 > 0$.*

Let a and ψ be as in (2.98), then for all solutions u of (2.95)

$$\mathbb{P}\Big\{ \sup_{t \in [0, T_0 \varepsilon^{-2}]} \|P_c \mathrm{Res}(\varepsilon w)(t)\| \leq \ln(\varepsilon^{-1})\varepsilon^{2-2\kappa},$$

$$\sup_{t \in [0, T_0 \varepsilon^{-2}]} \|P_s \mathrm{Res}(\varepsilon w)(t)\| \leq \ln(\varepsilon^{-1})\varepsilon^{3-\kappa} \Big\}$$

$$\geq 1 - \mathbb{P}\big\{ \|u(0)\| > \delta\varepsilon \big\} - \mathbb{P}\big\{ \|P_s u(0)\| > \delta\varepsilon^2 \big\} - o_\varepsilon(1) . \tag{2.101}$$

The main approximation result is the following. It is proved as Theorem 2.15 in [Blö05a]. We give some comments on the proof at the end of this section.

Theorem 2.13 **(Approximation)** *Suppose Assumptions 2.1, 2.4, and 2.9 are true. Fix $\delta > 0$, some small $1 \gg \kappa \geq 0$, and some time $T_0 > 0$. Let a and ψ be as in (2.98) and (2.99), then for all solutions u of (2.95)*

$$\mathbb{P}\left\{ \sup_{t \in [0, T_0 \varepsilon^{-2}]} \|u(t) - \varepsilon w(t)\| \leq \ln(\varepsilon^{-1})\varepsilon^{2-2\kappa} \right\}$$

$$\geq 1 - \mathbb{P}\left\{ \|u(0)\| > \delta\varepsilon \right\} - \mathbb{P}\left\{ \|P_s u(0)\| > \delta\varepsilon^2 \right\} - o_\varepsilon(1) .$$

We can for instance, apply the attractivity (cf. the more general Theorem 3.1 of [Blö05a]) on $[0, t_\varepsilon]$ and the approximation (cf. the more general result of Theorem 5.1 of [Blö05a]) on $[t_\varepsilon, T_0 \varepsilon^{-2}]$. Note we obviously cannot combine the versions of this section directly. Nevertheless, we derive

$$\mathbb{P}\left\{ \sup_{t \in [t_\varepsilon, T_0 \varepsilon^{-2}]} \|u(t) - \varepsilon w(t - t_\varepsilon)\| \leq \ln(\varepsilon^{-1})\varepsilon^{2-2\kappa} \right\}$$

$$\geq \mathbb{P}\left\{ \|u(0)\| \leq \delta\varepsilon \right\} - o_\varepsilon(1) ,$$

where $w(t)$ is defined in (2.98) with

$$a(0) = \varepsilon^{-1} P_c u(t_\varepsilon) \qquad \text{and} \qquad \psi(0) = \varepsilon^{-2} P_s u(t_\varepsilon) .$$

Let us finally remark that we can give estimates for the stopping time t^* from Assumption 2.10, as $t^* \geq T_0 \varepsilon^{-2}$ with high probability. This is interesting especially, as it is not yet settled, if our examples like surface growth or Bénard's problem exhibit unique global solutions, or not. Here we obtain, that the unique local solution does exist for a very long time with high probability.

The general outline for the proofs of Theorems 2.12 and 2.13 is similar to the proof of the residual and the approximation results in Theorem 2.5 and 2.6. However, there we were able to use cut-off techniques and Itô's formula. But in this case the technical tools used in the proof are quite different. We rely on path-wise estimates for stochastic integrals in Hölder norms using fractional integration by parts (see e.g. [Zäh98]). This is the main reason, why we need to assume trace-class noise in Assumption 2.9, as we need bounds on the paths of W in X.

Chapter 3

Applications — Some Examples

This chapter presents applications of the approximation via amplitude equations. The main results are about long-time behaviour of SPDEs or transient pattern formation for SPDEs on bounded domains. For simplicity of presentation, we focus on a few examples, in order to highlight the key ideas.

By no means we give an exhaustive presentation of all results possible, but focus on three examples. First we treat approximation of invariant measures near a change of stability. This is a review of results of [BH04]. We give the main ideas without stating details of the proofs.

The second section on pattern formation below threshold of instability gives a self-contained introduction, by explaining ideas and giving all proofs in the simplest setting possible. The final section on approximative centre manifolds and approximation of random attractors gives only the main ideas of proofs.

Invariant Measures

Section 3.1 states the approximation of invariant measures for the corresponding dual Markov semigroup. We summarise some of the results of [BH04]. Near the change of stability the invariant measure is well described in first order of ε by the invariant measure of the amplitude equation plus in second order by an infinite dimensional Ornstein-Uhlenbeck measure on the stable modes $\mathcal{S}$. The result is of the type $\mathbb{P}^{u^*} = \mathbb{P}^{\varepsilon a^*} \otimes \mathbb{P}^{\varepsilon^2 \psi^*}$. In this part the presentation is based on the setting and the results of Section 2.5. Apart from large deviation results, this is the first rigorous qualitative result for the structure of invariant measures for SPDEs with additive noise.

Another interesting application, that we nevertheless do not treat here, is the discussion of phenomenological bifurcation for SPDEs. It relies on the approximation of invariant measures. The invariant measure in $\mathcal{N}$ for the amplitude equation is usually easy to describe. For instance one can use the celebrated Fokker-Planck equation (cf. Risken [Ris84]), where we identify $\mathcal{N}$ with some $\mathbb{R}^n$ with $n \in \{1, 2\}$ for many examples. The Fokker-Planck equation is a deterministic PDE, which solution provides a smooth Lebesgue density of invariant measures on $\mathcal{N}$.

The second order correction in $\mathcal{S}$ is trivial to calculate, it is the invariant measure of an Ornstein-Uhlenbeck process, which is explicitly known. Let us point out that in our examples it is independent of the bifurcation parameter. We can now relate qualitative changes of the invariant measure of the SPDE to the shape of the invariant measure of the amplitude equation.

Recall that there is an infinite dimensional analogy of the Fokker-Planck equation (cf. [Cer01; DPZ02]), which is a PDE on an infinite dimensional domain. But from this equation it is not known how to derive qualitative information for the shape of the invariant measure, which could lead to a result on phenomenological bifurcation.

Pattern Formation

The second part of the applications are the results in Section 3.2 that deal with pattern formation below the threshold of instability. It gives results for the well known experimental fact that one can see dominating pattern even below the change of stability, once the system is sufficiently close to the bifurcation.

A celebrated example can be found in [SA02; SR94; OdZSA04; OA03] for noise-induced convection rolls below the onset of convection. In this case the system is slightly below a change of stability in the unperturbed (deterministic) system, which undergoes on bounded domains a pitchfork bifurcation. The noise is usually additive space-time white noise, which is an idealised model of small random fluctuations induced by thermal noise.

The effect of pattern formation in the previously described experiments is mainly due to additive space-time noise softening the sharp transition of the bifurcation. The pattern are determined by the functions in the space $\mathcal{N}$, which dominates the dynamics. We also discuss the case of multiplicative noise. Here the pattern, once it appears, sustains for a long time. But in general, one cannot verify easily the appearance of the pattern, as multiplicative noise may allow the system to settle down to the trivial state. This is different for additive noise, where we actually can show that the pattern appears independently of the initial condition. Furthermore, it is visible for most of the times, but we cannot prove that it persists for all times. It might disappear frequently for a short period.

The pattern space $\mathcal{N}$ is not sufficient to explain modulated pattern observed in many models and experiments. See for example [Lyt96; LM99; MC00; Wal97; DEKS95] and many more. The dimension of $\mathcal{N}$ is usually finite and too small to explain complicated modulated pattern. We need the methods of Chapter 4 in order to get a high dimensional pattern space, that would explain these effects (cf. Figure 4.1). These modulated patterns are also present in the theory of spinodal decomposition of binary alloys. See the review article [BMPW05] or [BMPW01; BMPW07]. However, the dynamics are completely different, as models for spinodal decomposition are already far in the unstable regime.

Approximative Centre Manifold

The third application in Section 3.3 is a result on approximative centre manifolds. The solutions are attracted and stay near the affine space $\mathcal{N} + \varepsilon^2 \psi^\star(t)$, where $\psi^\star$ is a stationary Ornstein-Uhlenbeck process in $\mathcal{S}$, while $\mathcal{N}$ is as usual the kernel of the linear operator L. The flow on the affine space is then on the much slower time-scale and given by the amplitude equation. We present results of [BH05], which also provide an approximation of random fixed points. Additionally, we sketch in Section 3.3.2 how to extend these results to the approximation of random attractors for the random dynamical system generated by the SPDE.

There are a lot more interesting problems where the approximation via amplitude equations applies. For instance we could lift from SDEs to SPDEs results on dynamic bifurcations, where a bifurcation parameter is time dependent and slowly driven through a bifurcation point. This was already briefly indicated in [Blö03]. The problem is especially important, as this method is a common tool used by experimentalists to derive bifurcation diagrams. The control parameter is slowly changed, and the assumption is that the system settles down sufficiently fast, such that one always observes stable stationary solutions. But it was already verified in [BG02a] for a pitchfork bifurcation in an SDE, that the stochastic forcing has a significant impact on the picture obtained by this method, at least in a small neighbourhood of the bifurcation point. There hysteresis type results are observed, although they are not present in the original equation.

3.1 Approximation of Invariant Measures

This section reviews the results obtained in [BH04] on approximating the invariant measure of SPDEs of the type of (1.2) near a change of stability. For simplicity the result is based on the setting of Section 2.5, where we considered a stable cubic nonlinearity and additive noise. To be more precise, consider equation (1.2) and let Assumptions 2.5, 2.7, and 2.8 be true. Additive noise is important, in order to have a unique exponential attracting invariant measure for the amplitude equation (cf. Assumption 3.1 and the discussion below).

It is a main issue to have the speed of convergence to the invariant measure for the amplitude equations under control (see (3.7)). The flow has to be (up to small errors) a contraction on the space of probability measures. This makes multiplicative noise more complicated, as there could be more than one invariant measure, and the speed of convergence is not controlled, as even nearby initial conditions may converge to different measures. A similar problem arises, when the amplitude equation is deterministic, for example, if the noise strength in the SPDE is $\mathcal{O}(\varepsilon^3)$. Here only partial results are available. Again problems arise with the speed of convergence in the amplitude equation, once its deterministic attractor is not trivial.

Before we turn to our main results let us fix some notation. For a general survey on invariant measures and Markov semigroups for SPDEs see for instance [DPZ96]. For the purpose of the presentation here, we define:

Definition 3.1   If solutions u of (1.2) are unique in law, then we denote by $\{\mathcal{P}_t^{(u)}\}_{t\geq 0}$ the dual Markov transition semigroup corresponding to (1.2), which means that $\mathcal{P}_t^{(u)}\mathbb{P}^{u(0)} = \mathbb{P}^{u(t)}$ is defined by the law of the solution $u(t)$ for all times $t \geq 0$.

We call a measure μ on the space X an *invariant measure* of the Markov semigroup, if $\mathcal{P}_t^{(u)}\mu = \mu$ for all $t > 0$.

Definition 3.2   We call a stochastic process $\{u(t)\}_{t\geq 0}$ in X a *stationary realization* of an invariant measure $\mu^\star$ on X if the law of $u(t)$ is $\mathbb{P}^{u(t)} = \mu^\star$ for all $t \geq 0$, and $u(t)$ is a mild solution in X of the SPDE (1.2). For short we say u is a stationary solution of (1.2).

The results of this section rely on a very mild non-degeneracy condition for the stochastic forcing defined in Assumption 2.8. This is the content of the next assumption. We remark without proof that this condition could be relaxed for some examples.

Assumption 3.1   *Consider $\mathcal{N}$ from Assumption 2.5 and W from Assumption 2.8. The covariance operator $P_c Q Q^* P_c$ of the noise on $\mathcal{N}$ is invertible as an operator from $\mathcal{N}$ to $\mathcal{N}$.*

Remark 3.1   *The key point of Assumption 3.1 is that the covariance operator of the Brownian motion in the amplitude equation has full rank. This ensures that the amplitude equation (1.5) has a unique invariant measure, which has an C^∞-density with respect to the Lebesgue measure. Moreover, we have exponential convergence of the laws of solutions towards the invariant measure. This relies on the dual Kolmogorov equation, which is also called Fokker-Planck equation. See e.g. [MT93; Cer01] or [Ris84]. See also [GM01] or [MS01] for related results in infinite dimensions.*

There are many situations where assumptions like 3.1 also ensure the existence of a unique invariant measure for the original SPDE. See for example the recent articles [KS01; Mat02; EL02; HM06] and the references therein.

Main Idea

The idea of approximating invariant measures is the following. Let $u^\star$ be a stationary solution of (1.2) and $a^\star$ a stationary solution of the amplitude equation. Consider $\mathbb{P}^{a^\star}$ as the law of $a^\star$ not only on $\mathcal{N}$, but extended on the full space X. Now

$$\mu^\star = \mathbb{P}^{u^\star} = \mathcal{P}_{T\varepsilon^{-2}}^{(u)}\mathbb{P}^{u^\star} \approx \mathbb{P}^{\varepsilon a(T)} \overset{T\to\infty}{\longrightarrow} \mathbb{P}^{\varepsilon a^\star} , \tag{3.1}$$

where a is a solution of the amplitude equation with initial condition $a(0) = \varepsilon^{-1} P_c u^\star$. The idea is to use the approximation result in the middle step of (3.1). But, as the approximation result is only valid on a finite time horizon $T \in [0, T_0]$, we need the precise convergence rate in order to control the last step in (3.1). Here we use the exponential convergence towards the unique invariant measure of the amplitude equation. Furthermore, we also take into account second order corrections of the invariant measure arising from the Ornstein-Uhlenbeck process, which gives the $\mathcal{O}(\varepsilon^2)$ correction.

Norms on Probability Measures

We make use of two different norms to measure the distances between invariant measures. See [Rac91] for a beautiful discussion of the relationship and properties between various metrics on probability measures.

Definition 3.3 The Wasserstein norm $\|\cdot\|_{\text{Lip}}$ (also called Kantorovich distance) is defined as the dual norm to the Lipschitz norm:

$$\|\phi\|_{\text{Lip}} = \sup_{x,y \in X} \left\{ |\phi(x)| , \; \frac{|\phi(x) - \phi(y)|}{\|x - y\|} \right\}, \qquad (3.2)$$

where $\phi : X \to \mathbb{R}$ is a Lipschitz continuous functions.

This means that for measures μ, ν on X

$$\|\mu - \nu\|_{\text{Lip}} = \sup_{\|\phi\|_{\text{Lip}}=1} \left(\int \phi \, d\mu - \int \phi \, d\nu \right). \qquad (3.3)$$

This norm implies convergence for probabilities. To be more precise:

Proposition 3.1 *For measures μ_n and μ we have that $\lim_{n \to \infty} \|\mu_n - \mu\|_{\text{Lip}} = 0$ implies $\lim_{n \to \infty} \mu_n(M) = \mu(M)$ for all measurable sets M.*

Additionally, if one has uniform bounds on arbitrary moments as in Theorem 2.8, then one also obtains convergence of moments and other statistical quantities.

Definition 3.4 The total variation norm $\|\cdot\|_{\text{TV}}$ is defined as the dual norm to the L^∞-norm.

Since $\|\phi\|_{\text{Lip}} \geq \|\phi\|_\infty$, the total variation norm is stronger than the Wasserstein norm. Note that the Wasserstein norm depends strongly on the metric that equips the underlying space, whereas the total variation norm is independent of that metric.

Example 3.1 The Wasserstein norm between two Dirac measures δ_x and δ_y is given by $\min\{1, \|x - y\|\}$, whereas $\|\delta_x - \delta_y\|_{\text{TV}}$ is 1 if $x \neq y$ and 0 otherwise.

3.1.1   The Results

Before we state our results, we introduce one more notation. For simplicity of presentation, we rescale the solutions of (1.2) by ε such that they are concentrated on a set of order 1 instead of a set of order ε. Furthermore, we rescale the equation to the slow time-scale $T = t\varepsilon^2$. Thus we consider v given by $v(T) = \varepsilon^{-1}u(T\varepsilon^{-2})$, where we split as usual $v = v_c + v_s$ ($v_c \in \mathcal{N}$, $v_s \in \mathcal{S}$). We obtain

$$
\begin{aligned}
\partial_T v_c &= & A_c(v_c + v_s) + \mathcal{F}_c(v_c + v_s) + \partial_T \beta \\
\partial_T v_s &= \varepsilon^{-2}Lv_s + A_s(v_c + v_s) + \mathcal{F}_s(v_c + v_s) + \partial_T \hat{W}_s ,
\end{aligned}
\tag{3.4}
$$

where $\hat{W}_s(T) = \varepsilon P_s QW(\varepsilon^{-2}T)$ and $\beta(T) = \varepsilon P_c QW(\varepsilon^{-2}T)$, as usual.

We denote by $\mu_\star^\varepsilon$ an invariant measure of (3.4). Note that the existence is standard using the celebrated Krylov-Bogoliubov method (cf. [DPZ96]).

Definition 3.5   Denote by $\nu_\star^\varepsilon$ the invariant measure for the pair of processes $(a, \varepsilon\psi)$, where the evolution is given by (1.5) and (1.6). Hence, in the slow time variable

$$
\begin{aligned}
\partial_T a &= A_c a + \mathcal{F}_c(a) + \partial_T \beta \\
\partial_T \psi &= \varepsilon^{-2}L\psi + \partial_T \hat{W}_s .
\end{aligned}
\tag{3.5}
$$

Denote by $\nu_\star^c$ the marginal on $\mathcal{N}$, and by $\nu_\star^s$ the one on $\mathcal{S}$, respectively.

Note that we actually do not need the uniqueness of $\nu_\star^\varepsilon$. We only need that the marginals on $\mathcal{N}$ and $\mathcal{S}$ are unique. The uniqueness of $\nu_\star^s$ is obvious, as we have an Ornstein-Uhlenbeck process. Furthermore, the uniqueness of $\nu_\star^c$ follows from Assumption 3.1.

Note that $\nu_\star^\varepsilon$ depends on ε by the rescaling of ψ. Recall also that we discussed in Remark 2.8 that the two noise terms in (3.5) may not be independent. Thus the equations in (3.5) are coupled through the noise, but actually they do not live on the same time scale, as the second equation in (3.5) lives on the fast time-scale t. However, as the equations are otherwise decoupled, we can determine the marginals $\nu_\star^c$ and $\nu_\star^s$ independently. The marginal $\nu_\star^c$ is independent of ε and $\nu_\star^s$ depends on ε only through the trivial scaling of $\varepsilon\psi$. Therefore we suppressed this ε-dependence in the notation.

With these notations, our main result in the Wasserstein distance is the following:

Theorem 3.1   (**Theorem 5.3 of [BH04]**) *Let Assumptions 2.5, 2.7, 2.8, and 3.1 hold. And let $\mu_\star^\varepsilon$ be any invariant measure of (3.4), and consider $\nu_\star^c$ and $\nu_\star^s$ from Definition 3.5.*

Then, for every $1 \gg \kappa > 0$ there is a constant $C > 0$ such that

$$
\|\mu_\star^\varepsilon - \nu_\star^c \otimes \nu_\star^s\|_{\mathrm{Lip}} \le C\varepsilon^{2-\kappa} .
\tag{3.6}
$$

Remark 3.2 *One can directly extend the idea of (3.1) to prove* $\|\mu_\star^\varepsilon - \nu_\star^\varepsilon\|_{\mathrm{Lip}} = \mathcal{O}(\varepsilon^{2-\kappa})$, *but the above formulation is more interesting, since* $\nu_\star^c$ *and* $\nu_\star^s$ *can be characterised explicitly, whereas* $\nu_\star^\varepsilon$ *can not. As mentioned already,* $\nu_\star^c$ *is given by the Fokker-Planck equation, while* $\nu_\star^s$ *is just an invariant Ornstein-Uhlenbeck measure, which is explicitly known. For a characterisation of* $\nu_\star^\varepsilon$ *the only hope is to rely on the independence of the two noise terms in (3.4), but this is in general not true, unless the covariance operator* Q *is block-diagonal with respect to the splitting* $X = \mathcal{N} \oplus \mathcal{S}$.

Idea of proof: Denote by $\{\mathcal{Q}_T\}_{T\geq0}$ the dual Markov transition semigroup (acting on measures) associated to (3.4), and by $\{\mathcal{P}_T\}_{T\geq0}$ the semigroup associated to the evolution of $(a(T), \varepsilon\psi(T))$. Then, the main ingredient for the proof of Theorem 3.1 is that there exists a time T_0 such that, for every pair (μ, ν) of probability measures with finite first moment, one has

$$\|\mathcal{P}_{T_0}\mu - \mathcal{P}_{T_0}\nu\|_{\mathrm{Lip}} \leq \frac{1}{2}\|\mu - \nu\|_{\mathrm{Lip}} + \varepsilon^2 \int_X (1 + \|P_s x\|)(\mu + \nu)(dx) . \tag{3.7}$$

This is exactly the contraction property of $\mathcal{P}_T$ mentioned after (3.1). The error term of order $\mathcal{O}(\varepsilon^2)$ is of technical nature, and arises from the fact, that we need to control uniformly exponential convergence rates that live on different time-scales. In order to prove (3.7), one uses the strong contraction properties of the linear dynamic in $\mathcal{S}$ and the strong mixing properties of the non-degenerate noise in $\mathcal{N}$, i.e. the exponential convergence rate to the unique invariant measure.

Once (3.7) is established, the proof of Theorem 3.1 follows in a rather straightforward way. One first obtains from the approximation result of Theorem 2.10 that

$$\|\mu_\star^\varepsilon - \nu_\star^\varepsilon\|_{\mathrm{Lip}} \leq \|\mathcal{Q}_{T_0}\mu_\star^\varepsilon - \mathcal{P}_{T_0}\mu_\star^\varepsilon\|_{\mathrm{Lip}} + \|\mathcal{P}_{T_0}\mu_\star^\varepsilon - \mathcal{P}_{T_0}\nu_\star^\varepsilon\|_{\mathrm{Lip}}$$
$$\leq \mathcal{O}(\varepsilon^{2-\kappa}) + \frac{1}{2}\|\mu_\star^\varepsilon - \nu_\star^\varepsilon\|_{\mathrm{Lip}} + \mathcal{O}(\varepsilon^2) ,$$

and therefore $\|\mu_\star^\varepsilon - \nu_\star^\varepsilon\|_{\mathrm{Lip}} = \mathcal{O}(\varepsilon^{2-\kappa})$. The bound $\|\nu_\star^\varepsilon - \nu_\star^c \otimes \nu_\star^s\|_{\mathrm{Lip}} = \mathcal{O}(\varepsilon^{2-\kappa})$ is then obtained by explicit calculations relying on the smoothness of the density of $\nu_\star^c$ with respect to the Lebesgue measure, combined with the separation of time scales between the dynamics on $\mathcal{N}$ and on $\mathcal{S}$. This is quite technical, and we refer the reader to [BH04] for details. □

In [BH04] we establish several additional results, which we briefly sketch here. The first result considers in the total variation norm only the marginals of the invariant measures on $\mathcal{N}$.

Theorem 3.2 **(Theorem 6.1 of [BH04])** *Suppose all Assumptions of Theorem 3.1 are true. Then for every* $\kappa > 0$ *there is a constant* $C > 0$ *such that*

$$\|P_c^* \mu_\star^\varepsilon - \nu_\star^c\|_{\mathrm{TV}} \leq C\varepsilon^{\frac{3}{2}-\kappa} . \tag{3.8}$$

Here $P_c^* \mu$ denotes the law of the map P_c under the measure μ. In this case it is just the marginal of μ on $\mathcal{N}$.

Idea of proof: We combine the smoothing properties of $P_c^* \mathcal{P}_T P_c^*$ with Theorem 3.1 to show that

$$\|\mathcal{P}_T \mu_\star^\varepsilon - \mathcal{P}_T \nu_\star^\varepsilon\|_{\mathrm{TV}} \le \frac{C\varepsilon^{2-\kappa}}{\sqrt{T}} \quad \text{for all} \quad T \in (0,1]. \tag{3.9}$$

Then, Girsanov's theorem implies

$$\|\mathcal{Q}_T \mu_\star^\varepsilon - \mathcal{P}_T \mu_\star^\varepsilon\|_{\mathrm{TV}} \le C\varepsilon\sqrt{T} \quad \text{for all} \quad T \in (0,1]. \tag{3.10}$$

Combining both estimates and optimising for T yields the result.   □

Remark 3.3   (**Remark 5.3 of [BH05]**) *The bound (3.8) is not always optimal. For example, if L and A are self-adjoint, Q is the identity and $\mathcal{F}$ is the negative gradient of a potential V (i.e., $\mathcal{F} = -\nabla V$), then the rescaled invariant measure $\mu_\star^\varepsilon$ for (1.2) can formally be written as*

$$\mu_\star^\varepsilon(du) = \exp\left(\tfrac{1}{2}\langle u, Au\rangle - V(u)\right)\mu_0^\varepsilon(du), \tag{3.11}$$

where μ_0^ε is the product of the Gaussian measure with covariance $\varepsilon^2 L_s^{-1}$ on $\mathcal{S}$ and the Lebesgue measure on $\mathcal{N}$. This explicit expression allows one to show that the density of $P_c^* \mu_\star^\varepsilon$ has derivatives of all orders and that these derivatives are all of order 1. This knowledge can be combined with Theorem 3.1 to show that in this case $\|P_c^* \mu_\star^\varepsilon - \nu_\star^c\|_{\mathrm{TV}} = \mathcal{O}(\varepsilon^{2-\kappa})$. However, this argument fails completely if, for example, $P_s Q^* Q P_s = 0$.

Our last result on the convergence of the invariant measures of the amplitude equation measures the distance between $\mu_\star^\varepsilon$ and $\nu_\star^c \otimes \nu_\star^s$ in the total variation norm. This however requires to impose a much stronger non-degeneracy assumption on the noise. The following assumption ensures, that there is enough noise on the high modes. Usually assumptions like these ensure the existence of a unique invariant measure for the SPDE and, furthermore, the exponential convergence in total variation norm. See for instance [MS01].

Assumption 3.2   *Let α be as in Assumption 2.6. Suppose there exists a constant $\gamma_0 > 0$, such that for all $\gamma \in [0, \gamma_0]$ the operators $\mathcal{F} : (X^\gamma)^3 \to X^{\gamma-\alpha}$ and $A : X^\gamma \to X^{\gamma-\alpha}$ are continuous. Furthermore, the operator Q^{-1} is continuous from $X^{\gamma_0-\alpha}$ to X and for some $\tilde\alpha \in [0, \tfrac{1}{2})$ we have $\|(1-L)^{\gamma_0-\tilde\alpha}Q\|_{\mathrm{HS}(X)} < \infty$.*

Theorem 3.3   (**Theorem 6.9 of [BH04]**) *Let Assumptions 2.5, 2.7, 2.8, and 3.2 hold. Then, for every $\kappa > 0$ there is a constant $C > 0$ such that*

$$\|\mu_\star^\varepsilon - \nu_\star^c \otimes \nu_\star^s\|_{\mathrm{TV}} \le C\varepsilon^{1-\kappa}. \tag{3.12}$$

Idea of proof: We denote by $\{\hat{\mathcal{P}}_T\}_{T\geq0}$ the dual Markov semigroup associated to the linear system

$$d\varphi = \varepsilon^{-2} L\varphi \, dT + dQ\tilde{W}(T) , \qquad (3.13)$$

where $\tilde{W}$ is a rescaled version of W. It is then possible to show by Girsanov's theorem that

$$\|\hat{\mathcal{P}}_T\mu_\star^\varepsilon - \mu_\star^\varepsilon\|_{\mathrm{TV}} = \|\hat{\mathcal{P}}_T\mu_\star^\varepsilon - \mathcal{Q}_T\mu_\star^\varepsilon\|_{\mathrm{TV}}$$
$$\leq C\varepsilon^{-\kappa/2}\sqrt{T} + C\varepsilon . \qquad (3.14)$$

Furthermore, the fast relaxation of the $\mathcal{S}$-component of the solutions to (3.13) toward its equilibrium measure, combined with the fact that the $\mathcal{N}$-marginals of $\mu_\star^\varepsilon$ and of $\nu_\star^\varepsilon$ are close by Theorem 3.2, allows to show $\|\hat{\mathcal{P}}_T\mu_\star^\varepsilon - \nu_\star^c \otimes \nu_\star^s\|_{\mathrm{TV}} \leq C\varepsilon$, provided $T \gg \varepsilon^2$. The result then follows by choosing T of the order $\varepsilon^{2-\kappa}$. □

3.2 Pattern Formation Below Criticality

This section gives examples for results that rigorously verify pattern formation below threshold of stability. The main aim is to give a flavour of the ideas in a more elementary setting, where we can give all proofs. For simplicity of presentation, we focus on the setting of nonlinearly stable deterministic operators in equations like (1.2) given for instance by Assumptions 2.5 and 2.7 (or Assumptions 2.1 and 2.2). Although the method will apply for much larger classes of equations as presented here. It mainly relies on the approximation result together with an analysis of the amplitude equation.

We focus first on additive (Assumption 2.8) and later on multiplicative noise (for example (2.3) with Assumption 2.3), as the results are quite different. In case of additive noise, we can show that the pattern is observable for most times, although it disappears frequently, as the noise drives the solution to any place in function space. Especially, the probability of reaching 0 in the near future is always positive. But for multiplicative noise we show that the pattern persists for very long times, provided it is already there at initial times. This result is quite natural, as under our general assumptions, the solution may tend to 0 for times to ∞ and the pattern might disappear in the limit.

3.2.1 *Additive Noise*

Let us consider as an example the Swift-Hohenberg equation (cf. Section 1.3). The equation falls into the scope of the equations treated in the previous section:

$$\partial_t u(t) = Lu(t) + \nu\varepsilon^2 u(t) - u^3(t) + \varepsilon^2 \partial_t QW(t) \quad \text{for} \quad t > 0. \qquad (3.15)$$

Fig. 3.1 A few examples for pattern in a space with four degrees of freedom on the domain $[-2\pi, 2\pi]^2$. The pattern space is span$\{\sin(x),\ \cos(x),\ \sin(y),\ \cos(y)\}$. It is invariant under translation and rotation of multiples of $\frac{\pi}{2}$. Note that this kind of examples cannot explain modulated pattern on large domains, as the pattern here are always strictly 2π-periodic in both directions.

We can consider different boundary conditions. For instance, we can choose zero Dirichlet type boundary conditions on $[0, \pi]$ and initial condition $u_0 = 0$. The operator $L = -(\Delta + 1)^2$ is given as in Section 1.3, and ν is some bifurcation parameter. We will see that the pattern in our toy-model is just the sine representing the convection roll in the full problem. We have no further degree of freedom in the dominant space.

In general the pattern space is $\mathcal{N}$, the kernel of L. The space $\mathcal{N}$ is finite dimensional, and in a lot of examples it is actually only a one or two dimensional space. A basis of this space then characterises the pattern. For example, if we consider (3.15) with periodic boundary conditions on the square $[0, 2\pi]^2$, then $\mathcal{N}$ consists of linear combinations of $\sin(x)$, $\cos(x)$, $\sin(y)$, and $\cos(y)$. We see that we have more than one particular pattern (cf. Figure 3.1). Nevertheless this still does not explain complicated modulated pattern, as already mentioned in the beginning of Chapter 3.

For the one-dimensional unperturbed Swift-Hohenberg equation, for instance subject to Dirichlet boundary conditions, it is well know that it undergoes a pitch-fork bifurcation at $\nu = 0$. There the trivial solution 0 gets unstable in the direction of $\sin \in \mathcal{N}$. For $\nu < 0$ the solution $u \equiv 0$ is the only stable solution, and for $\nu > 0$ we end up with a stable pattern that is a small deformation of the sine. Verifying this result for the Swift-Hohenberg equation or more general equations is a lot of work but standard, using for instance the celebrated theory of Crandel and Rabinowitz. See for example [Kie04].

In contrast to that, for the SPDE we will see that also in the case of $\nu < 0$ due to additive noise the pattern appears and it is visible for long times, although it should decay, due to the stability of $u = 0$. This is related to the well known fact that noise softens the sharp transition of the deterministic bifurcation. Recall that in our scaling the bifurcation parameter is of order of the noise strength away from the bifurcation.

In the following we verify a result like the probability of the pattern being visible for "most" $t \in [0, T_0 \varepsilon^{-2}]$ is near 1 for any $T_0 > 0$ and all $\varepsilon > 0$ sufficiently small. Let us start for simplicity at $u(0) = 0$, in order to avoid technical difficulties. If $u(0) \neq 0$, then simply use the attractivity result first. Now we apply the approximation result of Theorem 2.10, or 2.3 in case we are using multiplicative noise, in order to obtain that for all $p > 1$, $T_0 > 0$, and $\kappa > 0$, there is a constant $C > 0$ such that

$$\mathbb{E}\left(\sup_{t \in [0, T_0 \varepsilon^{-2}]} \|u(t) - \varepsilon a(\varepsilon^2 t)\|^p \right) \leq C \varepsilon^{2-\kappa} , \tag{3.16}$$

where a is a solution of the corresponding amplitude equation with $a(0) = 0$. Note that the constant C above is for the Swift-Hohenberg example actually independent of $\nu \in [-1, 1]$.

To prove a pattern result, we can for example verify that $|a(T)| \geq C \varepsilon^{1/2}$ for "a lot of" times $T \in [0, T_0]$. To quantify what we mean by most times, we first define

$$\mathcal{T}_\varepsilon(T) := |\{s \in [0, T] \ : \ |a(s)| \leq \varepsilon^{1/2}\}|$$

This defines the length of a stochastic set of "bad" times, where we possibly do not see the pattern. But we definitely see the pattern for a set of times of measure $(T_0 - \mathcal{T}_\varepsilon(T_0))\varepsilon^{-2}$. Note that the solution a of the amplitude equation (1.5) frequently crosses zero. Only for high dimensional $\mathcal{N}$ we could expect solutions to stay away from 0 with non-zero probability. This effect is similar to the well known Polya-theorem for Brownian motion.

To bound $\mathcal{T}_\varepsilon(T)$, it is easy to see that

$$\mathcal{T}_\varepsilon(T_0) = \int_0^{T_0} \chi_{\{|a(s)| \leq \varepsilon^{1/2}\}} ds .$$

As $\mathcal{T}_\varepsilon(T) \in [0, T]$ almost surely, we can bound arbitrary moments of $\mathcal{T}_\varepsilon(T_0)$. For instance, using Jensen's inequality, we derive for any $c > 0$

$$
\mathbb{E} \, e^{c \mathcal{T}_\varepsilon(T_0)} \leq \frac{1}{T_0} \int_0^{T_0} \mathbb{E} \exp\left(c T_0 \chi_{\{|a(s)| \leq \varepsilon^{1/2}\}}\right) ds
$$

$$
= \frac{1}{T_0} \int_0^{T_0} \mathbb{P}(|a(s)| > \varepsilon^{1/2}) ds + \frac{1}{T_0} e^{c T_0} \int_0^{T_0} \mathbb{P}(|a(s)| \leq \varepsilon^{1/2}) ds
$$

$$
= 1 + \frac{e^{c T_0} - 1}{T_0} \cdot \int_0^{T_0} \mathbb{P}(|a(s)| \leq \varepsilon^{1/2}) ds . \tag{3.17}
$$

Now it is obvious from $\mathbb{P}(|a(s)| = 0) = 0$ that

$$
p_\varepsilon := \frac{1}{T_0} \int_0^{T_0} \mathbb{P}(|a(s)| \leq \varepsilon^{1/2}) ds \longrightarrow 0 \quad \text{for} \quad \varepsilon \to 0 . \tag{3.18}
$$

It is crucial that (at least in law) a is independent of ε. See the discussion in Remark 1.2. Now for small $0 < \gamma_\varepsilon \ll 1$

$$
\mathbb{P}(\mathcal{T}_\varepsilon(T_0) \geq \gamma_\varepsilon) \leq \frac{e^{c T_0} - 1}{e^{c \gamma_\varepsilon} - 1} p_\varepsilon \to \frac{p_\varepsilon T_0}{\gamma_\varepsilon} \quad \text{for } c \to 0 .
$$

Defining $\gamma_\varepsilon = \sqrt{p_\varepsilon}$, we derive the following result.

Theorem 3.4   *Consider equation (1.2), with Assumptions 2.5, 2.7, and 2.8, and suppose for simplicity initial condition $u(0) = 0$. Then for any $T_0 > 0$*

$$
\mathbb{P}\left(\mathcal{T}_\varepsilon(T_0) \geq \sqrt{p_\varepsilon}\right) \leq T_0 \sqrt{p_\varepsilon} .
$$

Define for any $\kappa > 0$ the length of the set of times, where we see the pattern by:

$$
\mathcal{K}_\varepsilon(T_0) := \left| \left\{ t \in [0, T_0 \varepsilon^{-2}] : \; \|P_s u(t)\| \leq \varepsilon^{2-\kappa}, \; \|P_c u(t)\| \geq \varepsilon^{3/2} \right\} \right| .
$$

Then for all $p > 0$ there is a constant C depending on p, T_0, and κ, such that

$$
\mathbb{P}\left(\mathcal{K}_\varepsilon(T_0) \leq (T_0 - \sqrt{p_\varepsilon}) \varepsilon^{-2}\right) \leq T_0 \sqrt{p_\varepsilon} + C \varepsilon^p . \tag{3.19}
$$

The result in (3.19) states, that only for a small fraction of times we may not see the pattern. Actually, with high probability $\mathcal{K}_\varepsilon(T_0) \approx T_0 \varepsilon^{-2}$. The proof is straightforward using (3.16) and the bound on $\mathcal{T}_\varepsilon(T_0)$.

Remark 3.4   *We remark without proof that an analogy of Theorem 3.4 is straightforward for multiplicative noise. Use*

$$
\tilde{q}_\varepsilon = \frac{1}{T_0} \int_0^{T_0} \mathbb{P}(0 < |a(s)| \leq \varepsilon^{1/2}) ds \to 0 \quad \text{for } \varepsilon \to 0
$$

to bound

$$
\mathbb{P}(\mathcal{T}_\varepsilon(T_0) \geq \sqrt{\tilde{q}_\varepsilon}) \leq \mathbb{P}(a(0) = 0) + T_0 \sqrt{\tilde{q}_\varepsilon} .
$$

Nevertheless, in the next section, we prove a much stronger result. Here the main ingredient for the proof is $\mathbb{P}(a(s) = 0) = \mathbb{P}(a(0) = 0)$ *for all* $s > 0$, *which is easily verified via a transformation of the amplitude equation to a random ODE.*

At this stage it is essential, how good our bound on p_ε is. We already discussed that $p_\varepsilon = o_\varepsilon(1)$, but it is possible to give much better bounds for $\mathbb{P}(|a(s)| \leq \varepsilon^{1/2})$. One idea would be to control it via $\mathbb{E}|a(s)|^{-q}$ for $q > 0$. One problem is, that Itô's formula does not help much to bound the negative moment's. Furthermore, we expect the Lebesgue density of $\mathbb{P}^{a(s)}$ to be continuous and positive at 0. Thus negative moments of $|a(s)|$ do only exist for small $q > 0$.

Another idea is to use Birkhoff's ergodic theorem and the existence of exponentially attracting invariant densities for the amplitude equation. Here $p_\varepsilon \approx \mathbb{P}(|a^\star| \leq \varepsilon^{1/2})$ for T_0 large, where $a^\star$ is a stationary solution for the amplitude equation. We can then use the regularity of the invariant Lebesgue density of $\mathbb{P}^{a^\star}$ to explicitly determine the probability.

The last idea is to use regularity theory for $\mathbb{P}^{a(s)}$. It is enough to show that $\mathbb{P}^{a(s)}$ has a Lebesgue-density, which is around 0 uniformly bounded with respect to s. Then $p_\varepsilon \leq C\varepsilon^{1/2}$. It is also easy to see that this is the optimal bound, as in general we can not expect the density of $\mathbb{P}^{a(s)}$ being 0 at $a = 0$.

3.2.2 *Multiplicative Noise*

We already discussed a result for multiplicative noise in Remark 3.4. Here we actually show that the solution stays away from 0 for all $t \in [0, T_0\varepsilon^{-2}]$, once it is sufficiently far away. Obviously, $\mathbb{E}\sup_{t \in [0,T]} |a(t)|^{-2p} \leq C$ implies that $\mathbb{P}(a(t) \neq 0 \ \forall \ t \in [0,T]) = 1$. Our idea is to use Lemma B.11 to obtain bounds for negative moments, which in turn yield upper bounds on the probability that $a(t)$ is small.

For the first step let a be a solution of the amplitude equation (2.15) under the Assumption 2.2 or of (2.44) under the Assumption 2.4. Suppose furthermore $\mathbb{E}|a(0)|^{-4p} \leq \delta^2$. Now

$$\mathbb{P}(\mathcal{T}_\varepsilon(T_0) = 0) = \mathbb{P}\left(\inf_{t \in [0,T_0]} |a(t)| > \varepsilon^{1/2}\right)$$

$$\geq 1 - \varepsilon^p \mathbb{E}\left(\sup_{t \in [0,T_0]} |a(t)|^{-2p}\right)$$

$$\geq 1 - C\varepsilon^p \delta, \tag{3.20}$$

where we used Lemma B.11.

If $\mathbb{E}|a(0)|^{-4p}$ is large or infinite, then define $\tilde{a}$ to be a solution of the amplitude equation such that $a(0) = \tilde{a}(0)$ provided $|a(0)| \geq \delta_\varepsilon$ for some small $0 < \delta_\varepsilon \ll 1$. Otherwise we set $|\tilde{a}(0)| = 1$, arbitrary. Now it is easy to check that for all $p > 0$

$$\mathbb{E}|\tilde{a}(0)|^{-4p} \leq \mathbb{P}(|a(0)| \geq \delta_\varepsilon) \cdot \delta_\varepsilon^{-4p} + \mathbb{P}(|a(0)| < \delta_\varepsilon)$$

$$\leq 1 + \delta_\varepsilon^{-4p}. \tag{3.21}$$

Thus, by (3.20)

$$\mathbb{P}\left(\inf_{t \in [0,T_0]} |a(t)| \leq \varepsilon^{1/2} \right) \leq \mathbb{P}\left(|a(0)| \leq \delta_\varepsilon \right) + \mathbb{P}\left(\inf_{t \in [0,T_0]} |\tilde{a}(t)| \leq \varepsilon^{1/2} \right)$$

$$\leq \mathbb{P}\left(|a(0)| \leq \delta_\varepsilon \right) + C(1 + \delta_\varepsilon^{-4p})^{1/2} \cdot \varepsilon^p . \qquad (3.22)$$

Now we choose, for instance, $\delta_\varepsilon = \varepsilon^{1/2-\kappa}$ for $\frac{1}{2} > \kappa > 0$. We derive

$$\mathbb{P}\left(\inf_{t \in [0,T_0]} |a(t)| > \varepsilon^{1/2} \right) \geq \mathbb{P}(|a(0)| > \varepsilon^{\frac{1}{2}-\kappa}) - C\varepsilon^q \qquad (3.23)$$

for all $q > 0$, but with constant $C > 0$ depending on q and κ.

Let for simplicity u be a solution of (2.2) under Assumptions 2.1, 2.2, and 2.3. Then, we can bound the probability that we can see the pattern for a long time by

$$\mathbb{P}\left(\|P_c u(t)\| \geq \varepsilon^{3/2}, \ \|P_s u(t)\| \leq \varepsilon^{2-\kappa} \quad \forall t \in [0, T_0\varepsilon^{-2}] \right)$$

$$\geq \mathbb{P}\left(\sup_{t \in [0,T_0\varepsilon^{-2}]} \|u(t) - \varepsilon a(\varepsilon^2 t)\| \leq \varepsilon^{2-\kappa}, \ \inf_{t \in [0,T_0]} |a(t)| \geq \varepsilon^{1/2} \right) \qquad (3.24)$$

$$\geq \mathbb{P}(|a(0)| > \varepsilon^{\frac{1}{2}-\kappa}) - C\varepsilon^q - \mathbb{P}\left(\sup_{t \in [0,T_0\varepsilon^{-2}]} \|u(t) - \varepsilon a(\varepsilon^2 t)\| > \varepsilon^{2-\kappa} \right),$$

where $a(T)$ is as usual the solution of the amplitude equation with $a(0) = \varepsilon^{-1} P_c u(0)$. Using the approximation result of Theorem 2.3 we proved the following result.

Theorem 3.5   *Let u be a strong solution in X of (2.2) under Assumptions 2.1, 2.2, and 2.3. Suppose for some $\delta > 0$ and sufficiently large $p > 0$ we have $\mathbb{E}\|u(0)\|^{3p} \leq \delta\varepsilon^{3p}$ and $\mathbb{E}\|P_s u(0)\|^p \leq \delta\varepsilon^{3p}$.*

Then for all $T_0 > 0$, $\frac{1}{2} \gg \kappa > 0$, and $q > 0$ there is a constant $C > 0$ such that we can bound the probability of seeing the pattern for all times $t \in [0, T_0\varepsilon^{-2}]$ by

$$\mathbb{P}\left(\|P_c u(t)\| \geq \varepsilon^{3/2}, \ \|P_s u(t)\| \leq \varepsilon^{2-\kappa} \quad \forall t \in [0, T_0\varepsilon^{-2}] \right)$$

$$\geq \mathbb{P}(\|P_c u(0)\| > \varepsilon^{\frac{3}{2}-\kappa}) - C\varepsilon^q .$$

The condition for the moments of the initial condition $u(0)$ is exactly the one provided by the attractivity (cf. Theorem 2.1). This means it is always fulfilled, provided that we wait long enough. Any waiting time $T_w\varepsilon^{-2}$ with a small constant $T_w > 0$ is sufficient. Note finally, that we can prove an analogy of this theorem if we consider quadratic nonlinearities. Only $q > 0$ will be small in that case, as the approximation result is weaker.

3.3   Approximative Centre Manifold

This section describes how the evolution of solutions of a stochastic PDE subject to additive forcing is determined by an approximate centre manifold. This was briefly discussed in [Blö03] for the first order approximation. There the manifold is just the vector space $\mathcal{N}$. It attracts solutions up to errors of order $\mathcal{O}(\varepsilon^2)$, and the flow along $\mathcal{N}$ is given on the slow time-scale by the amplitude equation.

Here, we first state the results of Section 4.1 in [BH05], which relies on the second order correction introduced in [BH04] to describe the distance from $\mathcal{N}$, too. Therefore we need nonlinear stability, in order to bound moments. That is why we restrict ourselves in the following to nonlinear stable equations given by Assumption 2.7.

The key difference from results on random invariant manifolds (cf. for example [DLS03] or [MZZ07; DLS04; DW06b]) is that we obtain in first order $\mathcal{O}(\varepsilon)$ a fixed object, instead of a random set that is moving in time. Our result allows to control this dynamics at least to order $\mathcal{O}(\varepsilon^2)$ or $\mathcal{O}(\varepsilon^3)$. We pay for that qualitative description, by having all statements just with high probability, and not almost surely.

Our main result shows that in first order the flow of solutions of the SPDE (1.2) along $\mathcal{N}$ is well approximated by $\varepsilon a(\varepsilon^2 t)$, where a is the solution of the amplitude equation. In second order $\mathcal{O}(\varepsilon^2)$, the distance from $\mathcal{N}$ is given by fast oscillations, which is given as a stationary Ornstein–Uhlenbeck process $\varepsilon^2 \psi^\star(t)$. Thus the solutions are attracted by an $\mathcal{O}(\varepsilon^{3-\kappa})$-neighbourhood of $\varepsilon^2 \psi^\star(t) + \mathcal{N}$. Note that everything is valid only with high probability. Note that the setting for multiplicative noise is simpler, as the deterministic fixed point 0 is available. Therefore local results on the structure of invariant manifolds were obtained much earlier in that case (cf. for example [CLR01]). In Figure 3.2 the typical behaviour of solutions is given.

Theorem 3.6 *Suppose Assumptions 2.5, 2.7, and 2.8 are true, denote by $u(t)$ the mild solution in X of (1.2) with (random) initial condition u_0, and let $a(T)$ be the solution of (1.5) with $a(T_a) = \varepsilon^{-1} P_c u(T_a \varepsilon^{-2})$, where we fixed $T_a > 0$, arbitrary. Furthermore let $\psi^\star(t)$ be the stationary Ornstein–Uhlenbeck process solving the second order correction (1.6) given by (3.25) below.*

Fix any time $T_0 > T_a > 0$, some small $\kappa \in (0,1)$, and $p \geq 1$. Then, there exists a constant $C > 0$, such that

$$\mathbb{E}\left(\sup_{t \in [T_a, T_0]\varepsilon^{-2}} \| u(t) - \varepsilon a(t\varepsilon^2) - \varepsilon^2 \psi^\star(t) \|_X^p \right)^{1/p} \leq C\varepsilon^{3-\kappa} .$$

Remark 3.5 *Let us remark that from the formulation of the attractivity of Theorem 2.8 it is obvious, that instead of $T_\varepsilon = T_a \varepsilon^{-2}$, we can consider $t_\varepsilon = \mathcal{O}(\ln(\varepsilon^{-1}))$, if we allow only for initial conditions of order $\mathcal{O}(\varepsilon)$. In this case, we only need the attractivity result based on the linear stability in $\mathcal{S}$, which takes place on time-scales of order $\mathcal{O}(\ln(\varepsilon^{-1}))$.*

Proof. First we use global nonlinear attractivity in time $t_\varepsilon^{(1)} = \mathcal{O}(\varepsilon^{-2})$ for arbitrary initial conditions (cf. Theorem 2.8). Then we approximate with solution $\tilde{a}(t)$ of (1.5) and $\tilde{\psi}(t)$ of (1.6) for times $t \in [t_\varepsilon^{(1)}, T_0 \varepsilon^{-2}]$. Note that we need to shift the

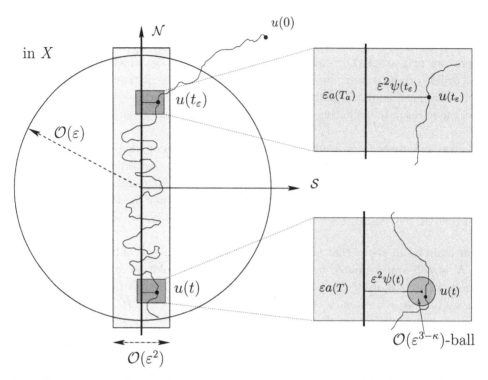

Fig. 3.2 Typical trajectory on the approximate centre manifold. Attractivity for times t between 0 and $t_\varepsilon = T_a \varepsilon^{-2}$, and approximation between t_ε and $T_0 \varepsilon^{-2}$. This figure is similar to Figure 1 of [BH05]

time appropriately. Define a version of the stationary Ornstein-Uhlenbeck process by

$$\psi^\star(t) = \int_{-\infty}^t e^{-L(t-s)} \, dP_s QW_*(s) \, , \tag{3.25}$$

where $W_*(s) = W(s)$ for $s > 0$ and $\{W_*(s)\}_{s \in \mathbb{R}}$ is a two-sided Wiener process. For the Brownian motion β in the amplitude equation, we need only the rescaling $\beta(T) = \varepsilon P_c Q W(T \varepsilon^{-2})$ for $T > 0$.

The difference between $\tilde{\psi}$ and $\psi^\star$ is trivially small in any p-th moment, if we wait for a sufficiently large logarithmic time $t_\varepsilon^{(2)}$. Define now $t_\varepsilon := t_\varepsilon^{(1)} + t_\varepsilon^{(2)}$, and choose $T_a > \varepsilon^2 t_\varepsilon$.

The difference between $\tilde{a}(t)$ and $a(t)$ is small by the approximation result, because first $\|\tilde{a}(t_\varepsilon) - a(t_\varepsilon)\| = \mathcal{O}(\varepsilon^{3-\kappa})$ by Theorem 2.10. Then, by the same theorem

$$\|\tilde{a}(t) - a(t)\| \leq \|\tilde{a}(t) - P_c u(t)\| + \|P_c u(t) - a(t)\| = \mathcal{O}(\varepsilon^{3-\kappa})$$

uniformly in $t \in [T_a, T_0] \varepsilon^{-2}$. $\square$

3.3.1 *Random Fixed Points*

Let us discuss the dynamics of random fixed points for random dynamical systems induced by the SPDE. At this point we do not give a precise definition of random dynamical systems (see [Arn98] for details) or random fixed points (see for example [Sch98]). In this section it is enough to know that a random fixed point induces a stationary solution for the SPDE, if we start the SPDE in the random fixed point.

Theorem 3.7 *Suppose Assumptions 2.5, 2.7, and 2.8 are true, and let $u^*(t)$ be a stationary mild solution in X of (1.2). Let a be the solution of (1.5) with $a(0) = \varepsilon^{-1} P_c u^*(0)$. Furthermore let ψ^* be the stationary Ornstein–Uhlenbeck given by (3.25).*

Then there is a constant $c_0 > 0$ such that for all $T_0 > 0$ any small $\kappa \in (0,1)$, and all $p \geq 1$, there exists a constant $C > 0$, such that

$$\mathbb{E}\left(\sup_{t \in [c_0 \ln(1/\varepsilon), T_0/\varepsilon^2]} \|u^*(t) - \varepsilon a(t\varepsilon^2) - \varepsilon^2 \psi^*(t)\|_X^p \right)^{1/p} \leq C\varepsilon^{3-\kappa} .$$

This result is an extension of the result for invariant measures (cf. Theorem 3.1), as the law of the stationary solution is at a fixed time t exactly an invariant measure. Here we can control the time-evolution, too.

Idea of proof: First we start the approximation result with initial condition $u^*(-T_a \varepsilon^{-2})$ for some $T_a > 0$. This implies first for $t = 0$, but then due to stationarity for all t, that

$$(\mathbb{E}\|u^*(t)\|^p)^{1/p} \leq C\varepsilon \qquad \text{and} \qquad (\mathbb{E}\|P_s u^*(t)\|^p)^{1/p} \leq C\varepsilon^2 .$$

Thus, we can start the approximation result in 0. and get

$$\mathbb{E}\left(\sup_{t \in [0, T_0/\varepsilon^2]} \|u^*(t) - \varepsilon a(t\varepsilon^2) - \varepsilon^2 \psi(t)\|_X^p \right)^{1/p} \leq C\varepsilon^{3-\kappa} ,$$

where ψ is the OU-process with initial condition $\psi(0) = \varepsilon^{-2} P_s u^*(0)$.

After a time scale of oder $\mathcal{O}(\ln(1/\varepsilon))$ we can approximate ψ with ψ^* as

$$\psi(t) = e^{tL}(\psi(0) - \psi^*(0)) + \psi^*(t) .$$

$\square$

Up to now, we do not use uniformity in the initial condition. Thus we can only prove results for random fixed points, and not for random set attractors. Nevertheless, the restriction to random fixed points still covers several cases. For example consider (1.2) with the nonlinearity $\varepsilon^2 \nu u - u^3$ for $\nu \in [-1,1]$. For $\nu < 0$ it is well known that the random attractor is just a single random fixed point, as it is easy to show via standard a priori estimates that all solutions attract each other exponentially fast.

But at $\nu = 0$ the stability changes. For $\nu > 0$ it is for a lot of linear operators L (for example $L = -(c + \Delta)^2$) completely open what the topology of the random attractor is. One exception being monotone SPDEs (see [Chu02]), where we for instance rely on maximum principles, in order to show that the attractor is only a single fixed point. See for example [CS04] and the detailed discussion in the introduction. Another exception are equations with Lipschitz continuous nonlinearities (cf. [CKS04]). But, nevertheless, in many examples of non-trivial random attractors for SPDEs, these attractors contain random fixed points.

3.3.2 *Random Set Attractors*

Let us extend our result for random fixed points (cf. Section 3.3.1) to random attractors. We do not present the result in full detail, but rather focus on a brief description of all steps necessary. Most steps are just quite technical but straightforward extensions of the estimates necessary for the residual, attractivity, and approximation. The key point is to rely on path-wise estimates, and take expectations in the end.

Assumption 3.3   *Consider eq. (1.2) fulfilling Assumptions 2.5, 2.7, and 2.8.*

The main example, we keep in mind is the stochastic Swift-Hohenberg equation in the space $X = L^2(G)$.

First we can use standard a priori estimates relying on nonlinear stability. This is very similar to Theorem 2.8, but nevertheless, we need to get uniform bounds with respect to the initial conditions

$$u(0) \in B_r := \{x \in X : \|x\| \leq r\}$$

for any fixed $r > 0$. For this we establish path-wise bounds for $v = u - \varepsilon^2 \phi$ with $\phi = W_{L-\varepsilon^2}$, which solves the following random PDE (compare (2.87))

$$\partial_t v = Lv + \varepsilon^2(Av + \phi) + \varepsilon^4 A\phi + \mathcal{F}(v + \varepsilon^2 \phi), \qquad v(0) = u(0). \qquad (3.26)$$

We use standard deterministic a priori estimates for (3.26), and take expectations in the end. Note that this transformation is not ergodic, in contrast to the usual transformation in the theory of random attractors, where one uses the stationary Ornstein-Uhlenbeck process for ϕ. For our setting we rely on ϕ, as we do not want to change initial conditions in (3.26).

The first step is the attractivity. It is similar to the proof of Theorem 2.8 and follows from standard a priori estimates for v and the stochastic convolution. Note that we rely on nonlinear stability. The result is that for all $r > 0$ there is a time $T_\varepsilon = \mathcal{O}(\varepsilon^{-2})$ such that for all $p > 0$

$$\mathbb{E}\left(\sup_{u(0) \in B_r} \|u(t)\|^p \right) \leq C\varepsilon^p \quad \text{for all} \ \ t \geq T_\varepsilon . \qquad (3.27)$$

As we use global nonlinear stability, we could take the whole space (i.e. $r = \infty$) to obtain the same result. Additionally, the second step of the attractivity yields that for some $t_\varepsilon = \mathcal{O}(\ln(\varepsilon^{-1}))$ we have

$$\mathbb{E}\left(\sup_{u(0)\in B_r} \|P_s u(t)\|^p\right) \leq C\varepsilon^{2p} \quad \text{for all} \quad t \geq T_\varepsilon + t_\varepsilon . \tag{3.28}$$

As we consider probabilities, we can equivalently use the pull-back convergence, where we start the solution in $u(-t)$ and evaluate the solution at $t = 0$, which is done for random attractors.

In the following we consider the random dynamical system generated by solutions of the SPDE in the following sense.

Definition 3.6 Under Assumption 3.3 define random maps

$$S(t,s) : X \to X$$

such that $S(t,s)u_0$ is the solution $\{u(t)\}_{t\geq s}$ for initial conditions $u(s) = u_0$, where we extend the driving Wiener process W two a two-sided Wiener process, as in the definition (3.25).

For simplicity of presentation, we do not use the usual notation of a cocycle

$$\varphi(t, \vartheta_s \omega) = S(t - s, 0; \vartheta_s \omega) = S(t, s; \omega)$$

for random dynamical systems generated by SPDEs, which involves the ergodic shift operator ϑ_s on the probability space. For details see [Arn98] or [Sch99].

Remark 3.6 *For the random dynamical system given by Definition 3.6 it is in many examples easy to show that there is a random set attractor. We will just assume that there exists one.*

For the precise definition and a detailed discussion of random attractors see also [CF94; CDF97; BCF93] or [Sch97; Sch99; FS96]. For the existence certain compactness properties for the deterministic flow are sufficient, in order to establish the existence of a random compact absorbing set for the stochastic system.

To prove our results it is enough to know the following definition, which is a slight abuse of notation, as a random set attractor has the property of this definition, but it is usually defined differently.

Definition 3.7 The random set attractor is a random compact set $\mathcal{A}_\varepsilon$ such that with probability one for all $r > 0$

$$\text{dist}(S(0, -t)B_r, \mathcal{A}_\varepsilon) \to 0 \quad \text{for} \quad t \to \infty . \tag{3.29}$$

Definition 3.8 The Hausdorff semi-distance is defined for sets $M, N \subset X$ by $\text{dist}(M, N) = \sup_{m \in M} \inf_{n \in N} \|m - n\|$. Thus $\text{dist}(M, N) = 0$ if and only if $M \subset N$.

Together with (3.27) and (3.28) one easily proves the following result:

Corollary 3.1 *Under Assumption 3.3 let $\mathcal{A}_\varepsilon$ be the random set attractor of the random dynamical system in X generated by the mild solution u of (1.2).*

Then for all $p > 0$ there is a constant $C > 0$ such that

$$\mathbb{P}(\mathcal{A}_\varepsilon \subset \mathcal{K}_\varepsilon) \geq 1 - C\varepsilon^p \,,$$

where for some suitable constants C_1, C_2 we defined the box

$$\mathcal{K}_\varepsilon = \{v \in X : \ \|P_c v\| \leq C_1 \varepsilon, \ \|P_s v\| \leq C_2 \varepsilon^2\} \,.$$

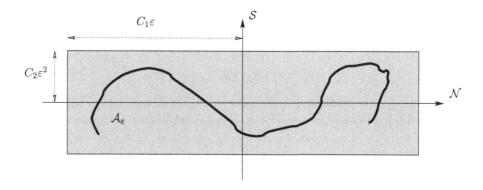

Fig. 3.3   First approximation for the random set attractor $\mathcal{A}_\varepsilon$. It is with high probability contained in the grey region.

The typical region given by the previous corollary containing the random set attractor is sketched in Figure 3.3. In the following, we show how to improve this result to the one sketched in Figure 3.4.

We do not give a detailed proof. The basic idea is to follow the proofs of the approximation result of [BH04], which is stated in Theorems 2.10 or 3.6. We need to establish most estimates path-wise for example from equation (3.26), and then take the supremum over the initial condition $u_0 \in B_r$ and expectation in the end. The generalisation is for most results straightforward. For instance, it is obvious how to achieve this for the bounds on a and ψ.

The major problem arise for the bound of the residual $P_c\text{Res}$. We will not give a detailed proof, as it is quite technical. Most of the terms in the residual are straightforward to handle, and we will comment only on the crucial term. Here we have to show that

$$\int_0^T B_a(\tau)\Big(e^{\tau L \varepsilon^{-2}} P_s u(0) + P_s W_L(\varepsilon^{-2}\tau)\Big) d\tau = \mathcal{O}(\varepsilon) \,,$$

where $B_a(\tau) := P_c\mathcal{F}(a(\tau), a(\tau), \cdot)$ is a linear operator, for a given solution a of the amplitude equation.

Let us comment on previous attempts to bound this term. In Lemma 4.7 of [BH04], we used Malliavin calculus, which could treat a quite general setting, but it is not helpful here. In [Blö05a] we used fractional integration by parts and Hölder estimates for a similar term. The simplest way is first to assume $\mathrm{tr}(Q^*Q) < \infty$, such that QW is a Wiener process in X. Secondly, assume that P_c commutes with Q. Then a is actually independent of P_sQW, and we can use a stochastic Fubini theorem together with integration by parts, in order to derive

$$\varepsilon \int_0^T B_a(\tau)P_sW_L(\varepsilon^{-2}\tau)d\tau$$
$$= \int_0^T \left[B_a(s) - \int_s^T B_a(\tau)(-L)\varepsilon^{-2}e^{(\tau-s)L\varepsilon^{-2}}d\tau \right] P_sQ\tilde{W}(s)ds\,,$$

where $\tilde{W}$ is a rescaled version of W. Now path-wise estimates for the residual are possible. However, we do not state the details here. We just wanted to motivate, why from now on we use trace-class noise and the fact that P_c commutes with Q.

The estimates for the approximation (cf. Section 4.3 of [BH04]) can be done by quite technical but standard path-wise bounds for the random PDE for the remainder $R = \varepsilon^{-2}(u - \varepsilon w)$, where as usual $\varepsilon w(t) = \varepsilon a(\varepsilon^2 t) + \varepsilon^2\psi(t)$. This relies heavily on the differentiability of $P_c\mathrm{Res}$ and nonlinear stability induced in Assumption 2.7 by (2.81) and (2.82).

Summarising everything yields the following result. We start the solution u at any time $-t \leq -2T_\varepsilon$ in $u(-t) \in B_r$, where T_ε was defined in (3.27). Now we apply first the modified attractivity, and furthermore the modified approximation on $[0, T_0\varepsilon^{-2}]$, to derive

$$\mathbb{E}\left(\sup_{u_0 \in B_r} \sup_{\tau \in [0,T_0/\varepsilon^2]} \|S(\tau,-t)u_0 - \varepsilon a(\varepsilon^2\tau) - \varepsilon^2\psi^\star(\tau)\|^p \right) \leq \varepsilon^{3p-\kappa} \quad \text{for all } t \geq 2T_\varepsilon\,.$$

This implies the following result, where we denote the time evolution of the random attractor $\mathcal{A}_\varepsilon(t) := S(t,0)\mathcal{A}_\varepsilon$. Actually the time evolution is only given by the shift operator on the probability space, i.e. $\mathcal{A}_\varepsilon(t;\omega) = \mathcal{A}_\varepsilon(0;\vartheta_t\omega)$.

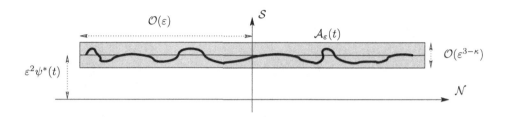

Fig. 3.4 Dynamics of the random attractor. The main idea to improve the result sketched in Figure 3.3 is the following. If we let the time evolve, then $P_s\mathcal{A}_\varepsilon(t)$ is uniformly attracted by the same stationary Ornstein-Uhlenbeck process $\varepsilon^2\psi^\star(t)$, at least with high probability.

Theorem 3.8 *Consider (1.2) with Assumptions 2.5, 2.7, and 2.8. Furthermore assume $\mathrm{tr}(Q^*Q) < \infty$ and that P_c commutes with Q. Let $\mathcal{A}_\varepsilon(t)$ be the time-evolution* *of the random set attractor of the random dynamical system in X generated by the mild solution u of (1.2). Moreover, let ψ^* be the stationary Ornstein-Uhlenbeck process defined in (3.25).*

Now for all $T_0 > 0$, $p \geq 1$, and $0 < \kappa \ll 1$ there are positive constant C_2 and C such that

$$\mathbb{P}\left(P_s \mathcal{A}_\varepsilon(t) \subset \varepsilon^2 \psi^*(t) + P_s B_{\varepsilon^{3-\kappa}} \text{ for all } t \in [0, T_0 \varepsilon^{-2}]\right) \geq 1 - C\varepsilon^p$$

and

$$\mathbb{P}(\|P_c \mathcal{A}_\varepsilon(t)\| \leq C_2 \varepsilon \text{ for all } t \in [0, T_0 \varepsilon^{-2}]) \geq 1 - C\varepsilon^p .$$

Thus with high probability, we know in which region the random attractor lies, and that it is oscillating in the $\mathcal{S}$-direction, as $\psi^*(t)$ is moving on the fast time-scale. This is also sketched in Figure 3.4.

Chapter 4

Amplitude Equations on Large Domains

4.1 Introduction

In this chapter we follow partly the presentation of [BHP05]. As already outlined in the introduction, this is the first rigorous approximation result by amplitude equations for SPDEs on large domains near a change of stability. We present a fairly general theorem for linear equations, but in order to keep notations at a reasonable level, we focus on approximating the stochastic Swift-Hohenberg equation by the stochastic (complex) Ginzburg-Landau equation, although the results do apply to larger classes of SPDEs or systems of SPDEs.

In the deterministic case similar results are well known for the Swift-Hohenberg equation. See for instance [CE90; KSM92; MSZ00]. However, there seems to be a lack of theory when noise is introduced into the system. Although this is frequently used on a formal level (see for example Section VI.D. of [CH93]). In particular, the treatment of extended systems (i.e., when the spatial variable takes values in an unbounded domain) is for SPDEs still out of reach of current techniques.

In a series of articles (for example [BMPS01; Blö03; BH04]) the amplitude of the dominating pattern was approximated by an SDE. This was already discussed in detail in Sections 1.1.1 or 2.5. But this approach shows its limitations on large domains, when the spectral gap in the linearised operator $\mathcal{L}$ (cf. ω in Assumption 2.1) between the dominating pattern and the remaining modes becomes small. These results are in particular not appropriate to explain modulated pattern occurring in many physical models and experiments (see for instance [Lyt96; LM99] or for a review [CH93]). See Figure 4.1 for an example of a one-dimensional modulated pattern.

The validity of the SDE-approximation is limited to a small neighbourhood of the stability change, which shrinks, as the size of the domain gets large. To be more precise, it fails, as soon as the second eigenvalue gets of order $\mathcal{O}(\varepsilon^2)$, which is the order of magnitude of the distance from the change of stability. In that case, we cannot control how many eigenvalues change sign. Hence, the spectral gap between the largest and the second eigenvalue determines the region of validity of the SDE approximation.

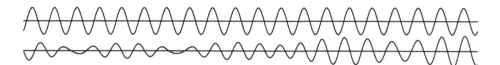

Fig. 4.1   An example of a modulated function on $[-\frac{\pi}{\varepsilon}, \frac{\pi}{\varepsilon}]$ with $\varepsilon = 0.05$. The upper part is 2π-periodic, while the lower part is a modulated function of the type $2\Re(\varepsilon A(\varepsilon x)e^{ix})$. Note that the modulated function is not even close to being 2π-periodic, but at least the zeros are approximately periodic.

For deterministic PDEs on unbounded domains it is well known (see for example [CE90; MS95; KSM92; Sch96]), that the dynamics of the slow modulations of the pattern can be described by an PDE which turns out to be of Ginzburg-Landau type. This was also indicated in the formal computation for SPDEs is Section 1.1.4.

Since the theory of translation invariant SPDEs on unbounded domains is still far from being fully developed, we adopt a somewhat intermediate approach, considering large but bounded domains of length $\mathcal{O}(\varepsilon^{-1})$, in order to avoid the technical difficulties arising for SPDEs on unbounded domains. Note that the same approach has been used in [MSZ00] to study the deterministic Swift-Hohenberg equation. In both the stochastic and deterministic approach the spectral gap of the linear operator $\mathcal{L}$ is of order $\mathcal{O}(\varepsilon^2)$. Furthermore, the eigenvalues are quite dense. When shifting the spectrum of $\mathcal{L}$ by an amount of order $\mathcal{O}(\varepsilon^2)$, $\mathcal{O}(\varepsilon^{-1})$-many eigenvalues may change sign. This is not the situation present on unbounded domains, where a whole band of uncountably many eigenvalues changes signs. Nevertheless, the number of eigenvalues in this large domain case goes to infinity, as $\varepsilon \to 0$. Furthermore, we will see that the choice of large but bounded domains captures and describes all the essential features of how noise in the original equation enters the amplitude equation.

Let us discuss the fact that it does not seem possible to adapt the deterministic theory to the stochastic equation. One major obstacle is that the whole theory for deterministic PDE relies heavily on good a priori bounds for the solutions of the amplitude equation in spaces of sufficiently smooth functions. Such bounds are unrealistic for our stochastic amplitude equation, as it turns out to be driven by space-time white noise. Its solutions are therefore at most α-Hölder continuous in space for $\alpha < 1/2$ and in time for $\alpha < 1/4$. Hence, one needs completely different techniques to estimate errors.

4.2   Setting

In this section we state rigorous results showing how to derive the stochastic Ginzburg-Landau equation as an amplitude equation for the stochastic Swift-Hohenberg equation. This justifies the formal calculation of Section 1.1.4. We focus on this simple example, but our results apply to larger classes of stochastic evolution equations. Roughly speaking all kinds of SPDEs or systems of SPDEs on the interval subject to periodic boundary conditions and with stable cubic non-linearities should be easy to treat without major changes in the proofs. We can also change the linear operator a lot. In Assumption 4.3 we define a very general operator $\mathcal{L}$.

Nevertheless, things may change drastically, if we change the boundary conditions. For example, if we consider the Swift-Hohenberg equation subject to Dirichlet or Neumann boundary conditions, then we have to modify the result, as we cannot use the complex notation. The amplitude equation would then be a real Ginzburg-Landau equation.

In the following we consider solutions to the one-dimensional stochastic Swift-Hohenberg equation

$$\partial_t U = -(1 + \partial_x^2)^2 U + \varepsilon^2 \nu U - U^3 + \varepsilon^{\frac{3}{2}} \xi_\varepsilon \,, \tag{4.1}$$

where $U(t, x) \in \mathbb{R}$ satisfies periodic boundary conditions for $x \in D_\varepsilon = [-L/\varepsilon, L/\varepsilon]$. The noise ξ_ε is assumed to be real-valued homogeneous space-time noise. To be more precise ξ_ε is a distribution-valued centred Gaussian field such that

$$\mathbb{E}\xi_\varepsilon(s, x)\xi_\varepsilon(t, y) = \delta(t - s)q_\varepsilon(|x - y|) \,. \tag{4.2}$$

For instance let ξ_ε be the generalised derivative of a Q_ε-Wiener process in $L^2(D_\varepsilon)$, where the covariance operator Q_ε is the convolution with $q_\varepsilon(|\cdot|)$. The family of correlation functions q_ε is assumed to converge in a suitable sense to a correlation function q. One should think for the moment of q_ε as simply being the $2L/\varepsilon$-periodic continuation of the restriction of q to D_ε. We state in Assumption 4.2 the precise assumptions on q and q_ε. This includes space-time white noise and noise with bounded correlation length, where q_ε is essentially independent of ε. See Figure 4.2 for an example.

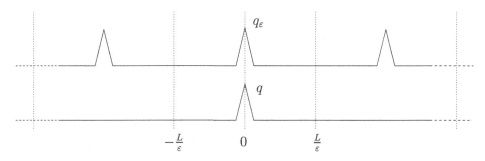

Fig. 4.2 An example for q and q_ε with bounded correlation length. Here $q(x) = 0$ for $|x| \geq \gamma$ and $q(x) = 1 - |x|/\gamma$ for $|x| \leq \gamma$. The correlation length is $2\gamma = |\mathrm{supp}\{q\}|$.

Note that we introduced a different scaling of the noise. On bounded domains the noise strength is always ε^2, in order to get an interesting stochastic approximation. This new scaling was already used in the formal calculation in Section 1.1.4, and is motivated by the fact, that we rescale not only time, but also space, and it assures that we get an ε-independent stochastic amplitude equation. If we consider noise-strength of order $\mathcal{O}(\varepsilon^2)$, then the amplitude equation is deterministic. This means, that on large domains, the distance from the bifurcation has to be much smaller than the noise strength, in order to see interesting stochastic effects.

We approximate solutions U of (4.1) as $U(t,x) \approx 2\varepsilon\Re(a(\varepsilon^2 t, \varepsilon x)e^{ix})$, where a is a solution of the following complex stochastic Ginzburg-Landau equation.

$$\partial_T a = 4\partial_X^2 a + \nu a - 3|a|^2 a + \sqrt{\hat{q}(1)}\,\eta\,, \quad X \in [-L, L]\,, \ a(0) = a_0\,, \qquad (4.3)$$

where η is complex space-time white noise. Here, a is subject to suitable ε-dependent boundary conditions. To be more precise those boundary conditions such that $a(T, X)e^{iX/\varepsilon}$ is $2L$-periodic in X.

Note that, as q is a correlation function, it is always non-negative semidefinite. Thus by definition the Fourier transform $\hat{q}(z) = \int_{-\infty}^{\infty} q(x)\exp(ixz)dx$ is always real and non-negative, and the noise strength in (4.3) is well defined.

Complex space-time white noise is for instance given by $\eta = (\xi_1 + i\xi_2)/\sqrt{2}$, where the ξ_k are independent copies of real-valued space-time white noise. Hence, η is a generalised Gaussian field such that

$$\mathbb{E}\eta(t,x) = 0, \quad \mathbb{E}\eta(t,x)\eta(s,y) = 0, \quad \text{and} \quad \mathbb{E}\eta(t,x)\overline{\eta(s,y)} = \delta(t-s)\delta(x-y)\,.$$

In order to simplify notations, we work from now on with the rescaled version $u(t,x)$ of the solutions of (4.1), defined through $U(t,x) = \varepsilon u(\varepsilon x, \varepsilon^2 t)$. Then, u satisfies the equation

$$\partial_t u = -\varepsilon^{-2}(1 + \varepsilon^2\partial_x^2)^2 u + \nu u - u^3 + \tilde{\xi}_\varepsilon\,, \qquad (4.4)$$

subject to periodic boundary conditions on the domain $D = [-L, L]$. Here, we define the rescaled noise $\tilde{\xi}_\varepsilon(t,x) := \varepsilon^{-3/2}\xi_\varepsilon(\varepsilon^{-1}x, \varepsilon^{-2}t)$. This is obviously a real-valued Gaussian noise with covariance given by

$$\mathbb{E}\tilde{\xi}_\varepsilon(t,x)\tilde{\xi}_\varepsilon(s,y) = \delta(t-s)\varepsilon^{-1}q_\varepsilon(\varepsilon^{-1}|x-y|)\,. \qquad (4.5)$$

If we have space-time white noise, then it is easy to see that this correlation function is (up to periodicity) in the bulk of the interval D completely independent of $\varepsilon > 0$.

We define the operator $\mathcal{L}_\varepsilon = -1 - \varepsilon^{-2}(1 + \varepsilon^2\partial_x^2)^2$ subject to periodic boundary conditions on $[-L, L]$, so that (4.4) can be rewritten as

$$\partial_t u = \mathcal{L}_\varepsilon u + (\nu + 1)u - u^3 + \tilde{\xi}_\varepsilon\,, \qquad (4.6)$$

subject to periodic boundary conditions on $[-L, L]$. In oder to handle the fact that the dominating modes $e^{\pm ix/\varepsilon}$ are not necessarily $2L$-periodic, we introduce the quantities

$$N_\varepsilon = \left[\frac{L}{\varepsilon\pi}\right], \quad \delta_\varepsilon = \frac{1}{\varepsilon} - \frac{\pi}{L}N_\varepsilon, \quad \text{and} \quad \rho_\varepsilon = N_\varepsilon\frac{\pi\varepsilon}{L}, \tag{4.7}$$

where $[z] \in \mathbb{Z}$ is used to denote the nearest integer of a real number z with the conventions that $[\frac{1}{2}] = 1$ and $[-z] = -[z]$. Note that $|\delta_\varepsilon|$ is therefore bounded by $\frac{\pi}{2L}$ independently of ε. Furthermore $\delta_\varepsilon = 0$ if and only if $L \in \varepsilon\pi\mathbb{N}$.

With these notations at hand, we rewrite the amplitude equation in a slightly different way. Defining $A(t,x) = a(t,x)e^{i\delta_\varepsilon x}$, (4.3) is equivalent to

$$\partial_t A = \Delta_\varepsilon A + (\nu+1)A - 3|A|^2 A + \sqrt{\hat{q}(1)}\eta, \quad \Delta_\varepsilon := -1 - 4(i\partial_x + \delta_\varepsilon)^2, \tag{4.8}$$

with *periodic* boundary conditions, where η is another version of complex space-time white noise. This transformation is purely for convenience, since periodic boundary conditions are more familiar. Furthermore, we see that all ε-dependent terms in the amplitude equation are uniformly small in L.

4.3   Approximation of the Stochastic Convolution

In this section, we give L^∞-bounds in time and in space on the difference between the stochastic convolutions of the original equation and of the amplitude equation. We use L^∞-norms in order to have the norm independent of a rescaling in the size of the domain. This is especially important, as we rescale the equations in space, in order to eliminate the ε-dependence of the spatial variable. Furthermore, as the Swift-Hohenberg equation describes an order parameter of the Rayleigh-Bénard problem, an L^∞-bound is even natural on the whole space, as we cannot expect decay conditions for solutions at infinity.

We define the stochastic convolutions $W_{\mathcal{L}_\varepsilon}(t)$ and $W_{\Delta_\varepsilon}(t)$, which are the solutions to the linearised equations, as

$$W_{\mathcal{L}_\varepsilon}(t) = \sqrt{Q_\varepsilon}\int_0^t e^{(t-\tau)\mathcal{L}_\varepsilon}\,dW_\xi(t) \tag{4.9}$$

and

$$W_{\Delta_\varepsilon}(t) = \sqrt{\hat{q}(1)}\int_0^t e^{(t-\tau)\Delta_\varepsilon}\,dW_\eta(t). \tag{4.10}$$

Here $\{W_\xi(t)\}_{t\geq 0}$ and $\{W_\eta(t)\}_{t\geq 0}$ denote standard cylindrical Wiener processes in $L^2(D)$ such that $\partial_t W_\xi$ and $\partial_t W_\eta$ are real- and complex-valued space-time white noise processes. Note that W_ξ is real valued, i.e. it is a cylindrical Wiener process in $L^2(D,\mathbb{R})$, while W_η is complex valued, i.e. in $L^2(D,\mathbb{C})$.

The definition of the covariance operator Q_ε is given in Definition 4.1 below and is such that the generalised process $\sqrt{Q_\varepsilon}\partial_t W_\xi$ has the covariance structure given in (4.5). We assume throughout the section that Assumption 4.2 holds for the correlation functions q and q_ε. In particular, note that Q_ε is a convolution operator and therefore commutes with the semigroup generated by $\mathcal{L}_\varepsilon$.

4.3.1 Noise

We now make precise the assumptions on the noise that drives our equation. We restrict ourself to correlation functions q in the following class:

Assumption 4.1   *The function $q : \mathbb{R} \to \mathbb{R}$ or distribution $q \in C_0^\infty(D)^*$ is such that for the Fourier transform $\hat{q}$ we have $\hat{q} \in L^\infty(\mathbb{R})$, $\hat{q} \geq 0$, and $\hat{q}$ globally Lipschitz continuous.*

At this point, a small technical difficulty arises from the fact that we want to replace ξ by a $2L/\varepsilon$-periodic translation invariant noise process ξ_ε which is close to ξ in the bulk of this interval. In the following, we present two examples how we can choose ξ_ε with (4.2). Denote by q^ε the $2L/\varepsilon$-periodic correlation function of ξ_ε and by q_k^ε its Fourier coefficients. This means

$$q_k^\varepsilon := \int_{-L/\varepsilon}^{L/\varepsilon} q^\varepsilon(x)\,e^{-ik\pi\varepsilon x/L}\,dx \quad \text{for} \quad k \in \mathbb{Z}. \tag{4.11}$$

One natural choice is to take for q^ε the periodic continuation of the restriction of q to $[-L/\varepsilon, L/\varepsilon]$. This does however not guarantee that q^ε is again positive semidefinite (i.e., $q_k^\varepsilon \geq 0$ for all $k \in \mathbb{Z}$). Another natural choice is to define q^ε via the Fourier coefficients of q by

$$q_k^\varepsilon := \hat{q}(k\pi\varepsilon/L) = \int_{-\infty}^\infty q(x)\,e^{-ik\pi\varepsilon x/L}\,dx \quad \text{for} \quad k \in \mathbb{Z}, \tag{4.12}$$

which corresponds to taking $q^\varepsilon(x) = \sum_{n=-\infty}^\infty q(x + 2nL/\varepsilon)$. This guarantees that q^ε is automatically positive semidefinite, but it requires some summability of q, which is especially a decay condition for $q(x)$ for $|x| \to \infty$. Note that for noise with bounded correlation length (i.e. support of q uniformly bounded) it is obvious, that (4.11) and (4.12) coincide for $\varepsilon > 0$ sufficiently small.

   We choose not to restrict ourselves to one or the other choice of q_ε, but to impose only a rate of convergence of the Fourier coefficients q_k^ε of q^ε towards $\hat{q}(k\pi\varepsilon/L)$:

Assumption 4.2   *Let q be as in Assumption 4.1. Suppose for all $\varepsilon > 0$ that there is a $2L/\varepsilon$-periodic function q_ε such that the corresponding Fourier coefficients q_k^ε (cf. 4.11) are a non-negative approximating sequence that satisfies*

$$\sup_{k \in \mathbb{Z}} |\sqrt{q_k^\varepsilon} - \sqrt{\hat{q}(k\pi\varepsilon/L)}| \leq C\varepsilon \tag{4.13}$$

for all sufficiently small $\varepsilon > 0$.

Example 4.1   A simple example of noise fulfilling Assumptions 4.1 and 4.2 is given by space-time white noise. Here $\hat{q}(k) = 1$ and the natural approximating sequence is $q_k^\varepsilon = 1$ for all $k \in \mathbb{Z}$.

Example 4.2 Another example fulfilling Assumptions 4.1 and 4.2 is a positive semidefinite function $q : \mathbb{R} \to \mathbb{R}$ such that $\mathrm{supp}(q) = \overline{\{x \in \mathbb{R} : q(x) \neq 0\}}$ is bounded. Then, obviously, $q_k^\varepsilon = \hat{q}(k\pi\varepsilon/L)$ for all $k \in \mathbb{Z}$ and $\varepsilon > 0$ sufficiently small.

For example a simple piece-wise linear hat function is sufficient (cf. Figure 4.2 or Section 5 of [Blö05b]).

A more general class of examples that allows for long-range correlations is given by the following lemma.

Lemma 4.1 **(Lemma 7.6 of [BHP05])** *Let q be positive semidefinite and such that $x \mapsto (1 + |x|^2)\, q(x)$ is in L^1. Define q_k^ε either by (4.12) or by (4.11) (in the latter case, we assume additionally that the resulting q^ε are positive semidefinite). Then Assumptions 4.1 and 4.2 are satisfied.*

Proof. This follows from elementary properties of Fourier transforms. □

One typical example in the class described in Lemma 4.1 is $q_\gamma(x) = e^{-|x|/\gamma}$, where we have an exponential decay rate, and all its restrictions q_γ^ε are again positive semidefinite. As the decay rate is sufficiently fast, one could call γ the typical correlation length, although this is not an example with bounded correlation length.

The covariance operator Q_ε in (4.9) is given by the following definition.

Definition 4.1 Let Assumption 4.2 be true. Then define Q_ε as the rescaled convolution with q^ε. This means

$$(Q_\varepsilon f)(x) = \frac{1}{\varepsilon} \int_{-L}^{L} f(y)\, q^\varepsilon \left(\frac{x-y}{\varepsilon} \right) dy . \tag{4.14}$$

On a formal level it is straightforward to verify that $\partial_t \sqrt{Q_\varepsilon} W_\xi$ is a noise with covariance structure as in (4.5).

4.3.2 *Main Result*

The main result of this section is:

Theorem 4.1 **(Theorem 7.1 of [BHP05])** *Let $W_{\mathcal{L}_\varepsilon}$ and W_{Δ_ε} be defined as in (4.9) and (4.10), and let the correlation functions q_ε with Fourier coefficients q_k^ε satisfy Assumption 4.2. Then for every $T > 0$, $\kappa > 0$, and $p \geq 1$ there exists a constant $C > 0$ and a joint realization of W_ξ and W_η such that*

$$\mathbb{E} \left(\sup_{t \in [0,T]} \sup_{x \in D} |W_{\mathcal{L}_\varepsilon}(t,x) - 2\Re(W_{\Delta_\varepsilon}(t,x)e^{ix})|^p \right) \leq C\varepsilon^{\frac{p}{2} - \kappa} \tag{4.15}$$

for every $\varepsilon \in (0,1)$.

We actually prove a more general result in Proposition 4.1 below, which has Theorem 4.1 as an immediate corollary. The general result allows a large class of linear operators $\mathcal{L}_\varepsilon$, instead of restricting it to the operator $-1 - \varepsilon^{-2}(1 + \varepsilon^2 \partial_x^2)^2$.

Idea of proof: The key idea is to use a Karhunen-Loeve type expansion of the stochastic convolution, which is basically a Fourier series expansion in time. The expansion in a Fourier series in $L^2(D)$ is standard for fixed $t > 0$, but we go further to expand in a series in $C^0([0, T] \times D)$ with complex-valued Gaussian coefficients. There is Lemma 4.2 that takes care of the L^∞ estimate.

The correlation between W_ξ and W_η is not an obvious rescaling like in the case of bounded domains. We have to specify the correlation of the Gaussian coefficients in the series expansion in the right way. It is especially important to get the correlation right for modes near the instability.   □

Lemma 4.2   (Lemma A.1 of [BHP05]) *Let* $\{\eta_k\}_{k \in I}$ *be i.i.d. standard Gaussian random variables (real or complex) with* $k \in I$ *an arbitrary countable index set. Moreover let* $\{f_k\}_{k \in I} \subset W^{1,\infty}(G, \mathbb{C})$ *where the domain* $G \subset \mathbb{R}^d$ *has sufficiently smooth boundary (e.g. piecewise* C^1 *). Suppose there is some* $\delta \in (0, 2)$ *such that*

$$S_1^2 = \sum_{k \in I} \|f_k\|_{L^\infty}^2 < \infty \quad \text{and} \quad S_2^2 = \sum_{k \in I} \|f_k\|_{L^\infty}^{2-\delta} \text{Lip}(f_k)^\delta < \infty$$

Define $f(\zeta) = \sum_{k \in I} \eta_k f_k(\zeta)$. *Then, with probability one,* $f(\zeta)$ *converges absolutely for any* $\zeta \in G$ *and, for any* $p > 0$, *there is a constant depending only on* p, δ, *and* G *such that*

$$\mathbb{E}\|f\|_{C^0(G)}^p \le C(S_1^p + S_2^p).$$

The proof of this Lemma is very similar to the proof of the celebrated Kolmogorov test, where one uses the embedding of Hölder spaces into fractional Sobolev spaces. We refrain from stating details here.

4.3.3   *Remarks*

Remark 4.1   (Higher Dimension) *The result of Theorem 4.1 or later Proposition 4.1 can not be generalised easily to dimensions higher than one, since the stochastic convolution* W_{Δ_ε} *of the Laplacian-type operator with space-time white noise is not even in* $L^2(D)$ *for domains* $D \subset \mathbb{R}^n$ *for* $n \ge 2$.

If the zeros of P, *the symbol of the differential operator* $\mathcal{L} = -P(i\partial_x)$, *are degenerate, which means that around all zeros* ζ_j $P(k) \approx (k - \zeta_j)^{2d}$ *for some* $d \in \{2, 3, \ldots\}$, *then we would obtain an amplitude equation with higher order differential operator, and we can proceed to higher dimension. The other option would be to use fractional noise in space, which is more regular than space-time white noise. Using the scaling invariance of fractional noise, we would obtain fractional noise in the amplitude equation.*

Remark 4.2 (**Space-Time White Noise**) *Note that we cannot avoid space-time white noise in the amplitude equation. Even if we consider noise with sufficiently fast decaying correlation function q. Our main results always show that we obtain space-time white noise in (4.8). Lemma 4.1 showed that a decay, where $x \mapsto (1 + |x|^2)q(x) \in L^1(\mathbb{R})$, is already sufficient.*

In all examples the stochastic convolution $W_{\mathcal{L}_\varepsilon}$ could be defined in $C^0([-L, L]^d)$ for all $d \in \mathbb{N}$, depending on the smoothness of q_ε, but we always end up with a second order PDE perturbed by space-time white noise that has solutions in $C^0([-L, L]^d)$ only for $d = 1$.

Remark 4.3 (**Localised Noise**) *We could also use noise which is not translation invariant. This could lead to an amplitude equation with point-noise. For example consider a cut-off function φ such that $\varphi(x) \to 0$ for $|x| \to \infty$. For the noise we use $\xi_\varepsilon^{(\varphi)}(t, x) = \varphi(x)\xi_\varepsilon(t, x)$, where ξ_ε is the usual noise discussed before. Now the correlation function of the rescaled $\tilde{\xi}_\varepsilon^{(\varphi)}$ is*

$$\mathbb{E}\tilde{\xi}_\varepsilon^{(\varphi)}(t, x)\tilde{\xi}_\varepsilon^{(\varphi)}(s, y) = \varepsilon^{-1}q^\varepsilon(|x - y|/\varepsilon)\varphi(x/\varepsilon)\varphi(y/\varepsilon)\delta(t - s) .$$

This a frequently used noise for equations on the whole real line, in order to use L^2-theory.

For the amplitude equation, we derive formally under suitable assumptions on φ that the limit for $\varepsilon \to 0$ is

$$\mathbb{E}\tilde{\xi}_\varepsilon^{(\varphi)}(t, x)\tilde{\xi}_\varepsilon^{(\varphi)}(s, y) \to \hat{q}(1)\delta(x)\delta(y)\varphi(0)^2\delta(t - s) .$$

Thus we end up with noise that acts only on the point 0. This means that a decay condition for the noise in (4.1) leads naturally to point-noise in 0 in the amplitude equation.

4.3.4   The General Result

We start by introducing the assumptions required for the differential operator $\mathcal{L} := -P(i\partial_x)$. See also Figure 4.3. The usual example being $P(\zeta) = (1 - \zeta^2)^2$ for the Swift-Hohenberg operator $-(1 + \partial_x^2)^2$.

Assumption 4.3 *Let P denote an even function $P : \mathbb{R} \to \mathbb{R}$ satisfying the following properties:*

P1 *P is three times continuously differentiable.*
P2 *$P(\zeta) \geq 0$ for all $\zeta \in \mathbb{R}$ and $P(0) > 0$.*
P3 *The set $\{\zeta \mid P(\zeta) = 0\}$ is finite and denoted by $\{\pm\zeta_1, \ldots, \pm\zeta_m\}$. Note that $\zeta_j \neq 0$.*
P4 *$P''(\zeta_j) > 0$ for $j = 1, \ldots, m$.*
P5 *There exists $r > 0$ such that $P(\zeta) \geq |\zeta|^2$ for all ζ with $|\zeta| \geq r$.*

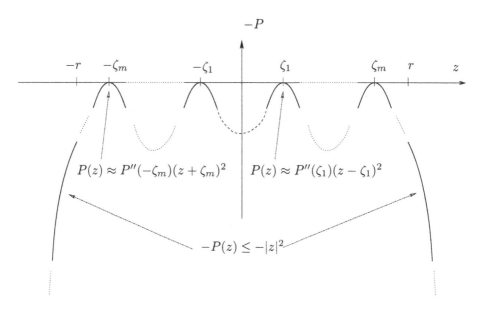

Fig. 4.3   A sketch of $-P$.

Note that choosing P even ensures that $-P(i\partial_x)$ is a real operator, but our results also hold for non-even P, up to trivial notational complications.

Let us now turn to the stochastic convolution $W_{\mathcal{L}_\varepsilon}$, which is the solution to the linear equation

$$dV(t) = \mathcal{L}_\varepsilon V(t)\,dt + \sqrt{Q_\varepsilon}\,dW_\xi(t)\,, \qquad (4.16)$$

where as before (cf. (4.6))

$$\mathcal{L}_\varepsilon = -1 - \varepsilon^{-2}P(\varepsilon i\partial_x)\,. \qquad (4.17)$$

Let us expand $W_{\mathcal{L}_\varepsilon}$ into a complex Fourier series. Denote as usual by $e_k(x) = e^{ik\pi x/L}/\sqrt{2L}$ the complex orthonormal Fourier basis in $L^2([-L,L],\mathbb{C})$. Define furthermore the symbol P^ε of the differential operator $\mathcal{L}_\varepsilon$ by

$$P^\varepsilon(k) = \frac{1}{\varepsilon^2}P\left(\frac{k\varepsilon\pi}{L}\right) + 1\,. \qquad (4.18)$$

Since Q_ε commutes with $\mathcal{L}_\varepsilon$, we can write the stochastic convolution as

$$
\begin{aligned}
W_{\mathcal{L}_\varepsilon}(t) &= \sqrt{Q_\varepsilon}\int_0^t e^{\mathcal{L}_\varepsilon(t-s)}dW_\xi(s)\\
&= \sum_{k=-\infty}^{\infty}\sqrt{q_k^\varepsilon}\,e_k\int_0^t \exp\{-P^\varepsilon(k)\,(t-s)\}\,dw_k(s)\,,
\end{aligned}
$$

where $W_\xi = \sum_{k=-\infty}^{\infty}\sqrt{q_k^\varepsilon}\,w_k e_k$ and the $\{w_k\}_{k\in\mathbb{Z}}$ are complex standard Wiener processes that are independent, except for the relation $w_{-k} = \overline{w_k}$. As only the modes near the dominant modes are important, we define the approximation of $W_{\mathcal{L}_\varepsilon}(t,x)$ by expanding P in a Taylor series up to order two around its zeroes. We thus define the approximating polynomials P_j^ε by

$$P_j^\varepsilon(k) = \frac{P''(\zeta_j)\pi^2}{2L^2}\left(k - \frac{L\zeta_j}{\varepsilon\pi}\right)^2 + 1 . \qquad (4.19)$$

At this point it is essential that $P''(\zeta_j) > 0$. With this notation, the approximation $\Phi = \Phi(t) \in \mathcal{C}^0(D)$ is defined by

$$\Phi(t) = 2\Re\left(\sum_{j=1}^m \sqrt{\hat{q}(\zeta_j)} \sum_{k=-\infty}^\infty e_k \int_0^t \exp\{-P_j^\varepsilon(k)(t-s)\}\, d\tilde{w}_{k,j}(s)\right), \qquad (4.20)$$

where the $\tilde{w}_{k,j}$'s are independent complex standard Wiener processes, to be fixed later during the proof. Furthermore, we need some correlation between the $\tilde{w}_{k,j}$'s and the w_k's.

The main approximation result is the following.

Proposition 4.1 **(Proposition 7.8 of [BHP05])** *Let Assumptions 4.3, 4.2 hold and consider $W_{\mathcal{L}_\varepsilon}$ and Φ as defined in (4.16) and (4.20). Then for every $T > 0$, $\kappa > 0$ and every $p \geq 1$, there exists a constant $C > 0$ and joint realizations of the processes W_ξ and η_i (i.e., of the processes w_k and $\tilde{w}_{k,j}$) such that*

$$\mathbb{E}\left(\sup_{x\in D} \sup_{t\in[0,T]} |\Phi(t,x) - W_{\mathcal{L}_\varepsilon}(t,x)|^p\right) \leq C\varepsilon^{p/2-\kappa} . \qquad (4.21)$$

The idea of proof was already sketched after Theorem 4.1. For details we refer the reader to [BHP05]. At this point, let us discuss a rewriting of Φ, in order to recover Theorem 4.1. We decompose $\frac{L\zeta_j}{\varepsilon\pi}$ into an integer part and a fractional part, so we write it as

$$\frac{L\zeta_j}{\varepsilon\pi} = \delta^{(j)} + k_j , \quad \delta^{(j)} \in \left[-\tfrac{1}{2}, \tfrac{1}{2}\right] , \quad \text{and} \quad k_j = \left[\frac{L\zeta_j}{\varepsilon\pi}\right] \in \mathbb{Z}. \qquad (4.22)$$

Here $[z]$ denotes the nearest integer to $z \in \mathbb{R}$ as in (4.7). Note that the scaling in L is slightly different than the one used in (4.7).

Define furthermore the projection

$$\pi_\varepsilon : L^2([-L,L],\mathbb{C}^m) \to L^2([-L,L],\mathbb{R})$$

$$A \mapsto 2\Re\left(\sum_{j=1}^m A_j(x)e^{i\pi k_j x/L}\right) .$$

With this notation, we can write Φ as $\Phi(t) = \pi_\varepsilon \Phi^a(t)$, where the j-th component of Φ^a solves the equation

$$\partial_t \Phi_j^a(t) = \Delta_j \Phi_j^a(t) + \sqrt{\hat{q}(\zeta_j)}\, \eta_j(t) , \qquad (4.23)$$

subject to periodic boundary conditions on $[-L,L]$. Here, the η_j's are independent complex-valued space-time white noise processes defined as the generalised derivative of the complex cylindrical Wiener processes $\sum_{k \in \mathbb{Z}} \tilde{w}_{k,j} e_k$. Finally, the Laplacian-type operator Δ_j is given by

$$\Delta_j = -\frac{P''(\zeta_j)}{2}\left(i\partial_x + \frac{\pi \delta^{(j)}}{L}\right)^2. \tag{4.24}$$

This operator is exactly the generalisation of Δ_ε, which was defined in (4.8). Note that in both cases, the operator is a perturbation of the Laplacian with terms uniformly small in L, as the coefficient $\pi \delta^{(j)}/L$ or δ_ε are bounded by $\pi/(2L)$.

4.4   Nonlinear Result

The main result of this section is the approximation result for solutions to (4.1) by means of solutions to the stochastic complex Ginzburg-Landau equation. We consider a class of "admissible" initial conditions given in the following Definition 4.2 below. It is a natural generalisation of the conditions needed for SPDEs on bounded domains. There admissible means that our initial condition is of order $\mathcal{O}(\varepsilon)$ on the dominating modes in $\mathcal{N}$, and $\mathcal{O}(\varepsilon^2)$ in $\mathcal{S} = \mathcal{N}^\perp$. Furthermore, the attractivity states that solutions are admissible after some time. Here the dimension of our space of dominating modes is of order $\mathcal{O}(1/\varepsilon)$. Thus it goes to infinity for $\varepsilon \to 0$. To control the solution on the dominant part, we need decay conditions on the Fourier modes, too.

Since we are dealing with a family of equations parameterised by $\varepsilon \in (0,1)$, we actually consider a family of initial conditions. We emphasise on the ε-dependence here, but we always consider it as implicit in the sequel.

Definition 4.2   A family of random variables A^ε with values in $L^2([-L,L],\mathbb{C})$ (or equivalently a family μ^ε of probability measures on $L^2([-L,L],\mathbb{C})$) is called *admissible* if there exists a decomposition $A^\varepsilon = W_0^\varepsilon + A_1^\varepsilon$, a constant $C_0 > 0$, and a family of positive constants $\{C_q\}_{q \geq 1}$ such that

1. $A_1^\varepsilon \in H^1([-L,L],\mathbb{C})$ almost surely and $\mathbb{E}\|A_1^\varepsilon\|_{H^1}^q \leq C_q$ for every $q \geq 1$,
2. The W_0^ε are centred Gaussian random variables such that

$$\left|\mathbb{E}\langle e_k, W_0^\varepsilon \rangle \langle e_\ell, W_0^\varepsilon \rangle\right| \leq C_0 \frac{\delta_{k\ell}}{1 + |k|^2}, \tag{4.25}$$

for all $k, \ell \in \mathbb{Z}$, ($\delta_{k\ell} = 1$ for $k = \ell$ and 0 otherwise).

A family of random variables u^ε with values in $L^2([-L,L],\mathbb{R})$ is called admissible if $u^+ e^{-iN_\varepsilon \pi x/L} \in L^2([-L,L],\mathbb{C})$ is admissible. Here, for $u = \sum_{k \in \mathbb{Z}} u_k e^{ik\pi x/L}$ we use $u^+ = \frac{1}{2}u_0 + \sum_{k=1}^\infty u_k e^{ik\pi x/L}$, which is the restriction of u to positive wave-numbers in Fourier space.

Remark 4.4 *Note that (4.25) implies that the covariance operator of W_0^ε commutes with the Laplacian, so that $W_0^\varepsilon = \sum_{k\in\mathbb{Z}} c_k^\varepsilon \xi_k e_k$ in law, where $0 \le c_k^\varepsilon \le C/(1+|k|)$ and the ξ_k are independent complex-valued standard normal random variables with the restriction that $\xi_{-k} = \overline{\xi_k}$. This implies by Lemma A.1 of [BHP05] that $\mathbb{E}\|W_0^\varepsilon\|_{C^0}^p \le C$ for every $p \ge 1$, as $\|e_k\|_{L^\infty} \le C$ and $\mathrm{Lip}(e_k) \le C|k|$, obviously.*

The concept of admissibility is basically a decay condition on the Fourier coefficients of solutions. The H^1-part decays sufficiently fast, while for the Gaussian part, we have a slow decay, as this part could fail to be even in $H^{1/2}$, but nevertheless, we get good bounds in C^0-spaces for the Gaussian part.

The key point here is the regularity theory of SPDEs. While deterministic parabolic PDEs exhibit exponentially fast decaying Fourier modes, the situation for SPDEs is different, as the noise prevents the modes from decaying fast. The H^1-part in the definition above reflects the deterministic smoothing, while the Gaussian part collects the effects of the noise.

It is a quite natural assumption to have admissible initial conditions, as solutions of the Swift-Hohenberg equation are admissible after some usually very large time. In particular the following result is true (cf. Theorem 1.1 or 5.1 of [BHP05]). As explained above, it balances the smoothing properties of the deterministic PDE with the roughening of the noise.

Theorem 4.2 **(Attractivity)** *Let u be a mild solution of (4.6) with arbitrary initial conditions in $L^2(D)$, then there is a deterministic time $t_\varepsilon > 0$ such that for all $t \ge t_\varepsilon$ the solution $u(t)$ is admissible in the sense of Definition 4.2.*

Our main result on the approximation is the following (cf. Theorem 1.2 or 4.1 of [BHP05]).

Theorem 4.3 **(Approximation)** *Let u be given by the mild solution of the rescaled Swift-Hohenberg equation (4.4) with an admissible initial condition written as $u_0(x) = 2\Re(A_0(x)e^{ix\rho_\varepsilon/\varepsilon})$. Consider the mild solution $A(T,X)$ to the stochastic Ginzburg-Landau equation (4.8) with initial condition A_0.*

Then, for every $T_0 > 0$, $\kappa > 0$, and $p \ge 1$, one can find joint realizations of the noises η and $\tilde{\xi}_\varepsilon$ such that

$$\left(\mathbb{E} \sup_{t\in[0,\frac{T_0}{\varepsilon^2}]} \sup_{x\in D} |u(t,x) - 2\Re(A(t,x)e^{ix\rho_\varepsilon/\varepsilon})|^p\right)^{1/p} \le C_{\kappa,p,T_0}\, \varepsilon^{1/2-\kappa} \qquad (4.26)$$

for every $\varepsilon \in (0,1]$.

Idea of proof: We dot state a detailed proof here. The general method is the same as sketched in Section 1.2, as we establish bound for the residual first. This mainly relies on the linear result of Theorem 4.1 and careful estimates for the mild formulation. Theorem 4.1 is also the main reason, why we loose an order $\varepsilon^{1/2}$ in (4.26).

For most of the approximation results, we use arguments similar to standard L^2-theory estimates. Nevertheless, we need to discuss the properties of the H^1-space with norm given by the quadratic form of $\mathcal{L}_\varepsilon$. It is essential that the embedding-constant of this space into L^p- or C^0-spaces is actually independent of ε.   □

Using an argument, which follows exactly the idea in (3.1), it is now straightforward to obtain an approximation result for invariant measures for (4.1) or (4.4) (see Theorem 6.1 of [BHP05]). The only technical problem is the bound on the exponential attractivity towards the unique invariant measure of (4.8). We need a uniform control, but it depends actually weakly on ε through δ_ε. But if we restrict ourselves to L's, where $\delta_\varepsilon = 0$, then the estimates are standard.

Let us fix some notations. The norm $\| \cdot \|_{\mathrm{Lip}}$ denotes the Wasserstein norm with respect to the C^0-topology for measures on $C^0([-L, L], \mathbb{R})$, while the mapping $\pi_\varepsilon : C^0([-L, L], \mathbb{C}) \to C^0([-L, L], \mathbb{R})$ is as usual defined by $(\pi_\varepsilon A)(x) = 2\Re(A(x)e^{ix\rho_\varepsilon/\varepsilon})$.

Theorem 4.4   (Invariant Measures) *Let $\nu_{\star,\varepsilon}$ be the invariant measure for the amplitude equation (4.8) and let $\mu_{\star,\varepsilon}$ be an invariant measure of the rescaled Swift-Hohenberg equation (4.4). Then, for every $\kappa > 0$ there is a constant $C > 0$ such that*

$$\|\mu_{\star,\varepsilon} - \pi_\varepsilon^* \nu_{\star,\varepsilon}\|_{\mathrm{Lip}} \leq C\varepsilon^{1/2-\kappa} \tag{4.27}$$

for every $\varepsilon \in (0, 1]$.

Let us remark that $\nu_{\star,\varepsilon}$ is actually independent of ε, provided $L \in \varepsilon\pi\mathbb{N}$. Furthermore, it should be straightforward to see that all measures $\nu_{\star,\varepsilon}$ are in an $\mathcal{O}(1/L)$-ball in Wasserstein distance, but we can never expect this error to be small in ε.

Appendix A

Basic Inequalities

In this appendix we present some basic results necessary for the proofs in Chapter 2. Most of these results are in principle well known, but we need slight modifications or more precise values of some constants.

First in Section A.1 we comment on the celebrated Burkholder-Davis-Gundy-type inequalities and maximal inequalities for stochastic integrals of convolution type. We present slight modifications and applications.

Next in Section A.2 we discuss a comparison argument for one-dimensional ODEs or SDEs. Lemma A.7 is an application that is helpful for determining the effects of nonlinear stability in a priori estimates. It shows that after some time solutions of certain ODEs can be uniformly bounded independent of the initial condition.

Before we proceed let us recall two celebrated results.

Theorem A.5   (Young's inequality)
For $p, q > 1$ such that $\frac{1}{q} + \frac{1}{p} = 1$ there is a constant $C > 0$ such that

$$xy \leq C(x^p + y^q) \qquad \text{for all } x, y > 0 \,.$$

Especially, for all $\varepsilon > 0$ there is a constant $C_\varepsilon > 0$ such that

$$xy \leq \varepsilon x^p + C_\varepsilon y^q \qquad \text{for all } x, y > 0 \,.$$

We also need the celebrated Itô Formula. We will state only a simplified version needed in this book. For the general case see for example [DPZ92].

Theorem A.6   (Itô's Formula)
Let $\{u(t)\}_{t \geq 0}$ be a stochastic process in some Hilbert-space X and let β be a standard real-valued Brownian motion. Suppose that

$$du = f(u)dt + g(u)d\beta$$

for some functions $g, f : X \to X$. Then for a twice continuously differentiable *function* $\varphi : X \to \mathbb{R}$, *we have*

$$\varphi(u(t)) - \varphi(u(0)) = \int_0^t D\varphi(u(s))[f(u(s))]ds + \int_0^t D\varphi(u(s))[g(u(s))]d\beta(s)$$
$$+ \frac{1}{2} \int_0^t D^2\varphi(u(s))[g(u(s)), g(u(s))]ds .$$

The most frequently used example for which we use Itô's formula is $\varphi(u) = \|u\|^p$ with

$$D\varphi(u)[v] = p\|u\|^{p-2}\langle u, v\rangle$$

and

$$D^2\varphi(u)[v_1, v_2] = p\|u\|^{p-2}\langle v_1, v_2\rangle + p(p-2)\|u\|^{p-4}\langle u, v_1\rangle\langle u, v_2\rangle .$$

A.1   Burkholder-Davis-Gundy Inequality

In this subsection, we discuss maximal inequalities of Burkholder-Davis-Gundy type, in order to bound stochastic integrals of convolution type. We consider only bounds for terms like $\mathbb{E}\sup_{t\in[0,T]} \left\| \int_0^t e^{(t-s)L} f(s)d\beta(s) \right\|^p$, where L is an operator given for instance by Assumption 2.1.

For a nice review of recent results see [HS01]. We will present some modifications needed for our applications. Especially we have to improve the dependence of the constant on the time T for large times. First we need the following well known theorems (see e.g. [DPZ96] or [HS01])

Theorem A.7   (Burkholder-Davis-Gundy) *Let β be a Brownian motion, and f some stochastic process adapted to β. Then for all $p > 0$ there is a constant $C > 0$ depending only on p such that*

$$\mathbb{E} \sup_{t\in[0,T]} \left\| \int_0^t f(s)d\beta(s) \right\|^p \leq C\mathbb{E}\left(\int_0^T \|f(s)\|^2 ds \right)^{p/2} . \tag{A.1}$$

A version, which is true for all martingales is the celebrated Doob inequality.

Theorem A.8   (Doob) *Consider f and β as in Theorem A.7. Then for arbitrary $p > 1$*

$$\mathbb{E} \sup_{t\in[0,T]} \left\| \int_0^t f(s)d\beta(s) \right\|^p \leq \frac{p}{p-1}\mathbb{E}\left\| \int_0^T f(s)d\beta(s) \right\|^p .$$

A simple lemma, which can also be based on a rescaling argument, is the following.

Lemma A.3 *Let β be a Brownian motion, and f some stochastic process adapted to β. Then there is a constant $C > 0$ independent of β and f such that*

$$\mathbb{E} \sup_{t \in [0, T_0 \varepsilon^{-2}]} \varepsilon \| \int_0^t f(\tau) d\beta(\tau) \| \leq C T_0^{1/2} \Big(\sup_{t \in [0, T_0 \varepsilon^{-2}]} \mathbb{E} \| f(t) \|^2 \Big)^{1/2}$$

for all $\varepsilon > 0$.

Proof. This is a direct consequence of Theorem A.7 and Hölder's inequality. $\square$

Note that the right hand side in (A.1) is easily bounded by $C T^{p/2}$ provided that $\mathbb{E} \sup_{t \in [0,T]} \| f(s) \|^p$ or $\sup_{t \in [0,T]} \mathbb{E} \| f(s) \|^p$ is bounded. Here we need to improve the dependence of the constant on $T > 0$, as in our applications the time $T = \mathcal{O}(\varepsilon^{-2})$ is typically very large.

Lemma A.4 *Fix $p > 4$. Let L be as in Assumption 2.1 and $\{\beta(t)\}_{t \geq 0}$ some Brownian motion. Then there is a constant $C > 0$ such that for all $T > 0$, $\delta > 0$, and all stochastic processes f adapted to β, we have*

$$\mathbb{E} \sup_{t \in [0,T]} \| \int_0^t e^{(t-s)L\delta} f(s) d\beta(s) \|^p \leq C T \delta^{1-p/2} \sup_{t \in [0,T]} \mathbb{E} \| f(s) \|^p .$$

Results of this type with a different T-dependence of the constant are well known even for bounded or contraction semigroups. See for example [HS01] or [Tub84]. Nevertheless, we establish here a much better control on the dependence of the constant on T and δ, which was not done before. Here we rely on exponential decay properties of the semigroup. For simplicity of presentation, we do not focus on optimal dependence, but nevertheless the lemma yields a very good result for large T and δ.

Proof. We use the celebrated factorisation method of Da Prato and Zabzcyck (cf. [DPZ92]). Based on a stochastic Fubini theorem,

$$I(t) := \int_0^t e^{(t-s)L\delta} f(s) d\beta(s) = C_\gamma \int_0^t e^{(t-s)L\delta} (t-s)^{\gamma-1} Y(s) ds , \tag{A.2}$$

for $\gamma \in (0,1)$ with some constant C_γ depending only on γ. Moreover,

$$Y(s) := \int_0^t e^{(t-s)L\delta} (t-s)^{-\gamma} f(s) d\beta(s) . \tag{A.3}$$

Fix γ such that $\frac{1}{p} < \gamma < \frac{1}{2} - \frac{1}{p}$. Using the stability of the semigroup from Assumption 2.1 and Hölder's inequality, we derive for $t \in [0, T]$

$$\|I(t)\|^p \leq C \Big(\int_0^t e^{-(t-s)\omega\delta}(t-s)^{\gamma-1}\|Y(s)\|ds \Big)^p$$

$$\leq C \Big(\int_0^t e^{-s\omega\delta\frac{p}{p-1}} s^{\frac{(\gamma-1)p}{p-1}} ds \Big)^{p-1} \int_0^t \|Y(s)\|^p ds$$

$$\leq C\delta^{1-\gamma p} \int_0^T \|Y(s)\|^p ds , \qquad (A.4)$$

where C denotes positive constants depending on p, but independent of T, δ, and f. Furthermore, using Lemma A.5

$$\mathbb{E}\|Y(t)\|^p \leq C\mathbb{E} \Big(\int_0^t \|e^{(t-s)L\delta}(t-s)^{-\gamma} f(s)\|^2 ds \Big)^{p/2}$$

$$\leq C\mathbb{E} \Big(\int_0^t e^{-2\omega(t-s)\delta}(t-s)^{-2\gamma}\|f(s)\|^2 ds \Big)^{p/2}$$

$$\leq C \Big(\int_0^t e^{-\omega s\delta\frac{p}{p-2}} s^{-\frac{2\gamma p}{p-2}} ds \Big)^{(p-2)/2} \cdot \int_0^t e^{-\omega(t-s)\delta p}\mathbb{E}\|f(s)\|^p ds$$

$$\leq \sup_{s\in[0,T]} \mathbb{E}\|f(s)\|^p \delta^{1+\gamma p-\frac{p}{2}} \cdot \delta^{-1} . \qquad (A.5)$$

Combining all results finishes the proof.   □

The following useful lemma is a moment inequality that generalises Itô's isometry. It follows directly from Theorem A.7, but it is much simpler, and usually used in the proof of Theorem A.7.

Lemma A.5   *For all $p > 0$ there is a constant C such that for all Hilbert-space valued stochastic processes f adapted to the filtration of the Brownian motion β*

$$\mathbb{E}\Big\| \int_0^t f(s)d\beta(s) \Big\|^p \leq C\mathbb{E} \Big(\int_0^t \|f(s)\|^2 ds \Big)^{p/2} .$$

See for example Section 1.7 of [Mao97] for a detailed discussion of moment inequalities and Burkholder-Davis-Gundy inequalities in an SDE setting.

A.2   Comparison Argument for ODEs

Results of this type are well known. See for example [Hal80]. Nevertheless, we state the result necessary for our applications in Lemma A.6 and give a simple proof. The application in Lemma A.7 will provide the key estimate to use nonlinear stability. In a simple ODE setting we show that after a fixed time, we can bound solutions completely independent of the initial condition. Let us first recall that $AC([0, T], \mathbb{R})$ denotes the space of all functions $f : [0, T] \rightarrow \mathbb{R}$, which are absolutely continuous.

Lemma A.6   *Let $f : [0, T] \times \mathbb{R} \to \mathbb{R}$ be some continuous function, such that f is locally Lipschitz uniform in t, i.e., for all intervals $[a, b] \subset \mathbb{R}$ there is a constant such that $|f(s, x) - f(s, y)| \leq C|x - y|$ for all $x, y \in [a, b]$ and all $s \in [0, T]$. Furthermore, let $x \in AC([0, T], \mathbb{R})$ such that $\dot{x}(t) \leq f(t, x(t))$ for all $t \in [0, T]$ and suppose that y is a solution of $\dot{y}(t) = f(t, y(t))$ on $[0, T]$.*

Then, if $x(0) \leq y(0)$ then $x(t) \leq y(t)$ for all $t \in [0, T]$.

Remark A.5   *The result remains true, even if we add additive noise to the equations. If we then consider the integrated equation, we can state a similar result.*

Proof.   Assume there is a time $t_0 > 0$ such that $x(t_0) > y(t_0)$. As x and y are continuous, there are times $0 < t_a < t_e \leq T$ such that $x(t_a) = y(t_a)$, and $x(t) > y(t)$ for $t \in (t_a, t_e]$. Due to the continuity of x and y there is an interval $[a, b]$ which contains $x(t)$ and $y(t)$ for all $t \in [t_a, t_e]$. Now, for all $t \in [t_a, t_e]$

$$0 \leq x(t) - y(t) = \int_{t_a}^{t} \dot{x}(s) - \dot{y}(s)ds \leq \int_{t_a}^{t} f(s, x(s)) - f(s, y(s))ds$$

$$\leq C \int_{t_a}^{t} |x(s) - y(s)|ds .$$

Finally, Gronwall's inequality (cf. Lemma A.8) yields $x(t) - y(t) = 0$ for all $t \in [t_a, t_e]$ contradicting the assumption. $\square$

A straightforward application is the following.

Lemma A.7   *Consider any absolutely continuous function $x : [0, T] \to [0, \infty)$. If $\dot{x} \leq \delta^2 - x^2$ for some $\delta > 0$ and all $t \in [0, T]$, then*

$$x(t) \leq \max\{2\delta, \frac{4}{3t}\} \quad \text{for all } t \in [0, T] .$$

Furthermore, if $\dot{x}(t) \leq \delta^{(p+2)/2} - x^{(2+p)/p}(t)$ for some $\delta > 0$, some $p \geq 2$ and all $t \in [0, T]$ then

$$x(t) \leq \max\left\{(2\delta)^{\frac{p}{2}}, \left(\frac{2p}{3t}\right)^{\frac{p}{2}}\right\} \quad \text{for all } t \in [0, T] .$$

Proof.   Consider the first assertion. If $x(t) \geq 2\delta$ then $\dot{x}(t) \leq -\frac{3}{4}x^2(t)$, then Lemma A.6 immediately yields

$$x(t) \leq \left(\frac{3}{4}t + \frac{1}{x(0)}\right)^{-1} \leq \frac{4}{3t}.$$

The second assertion is proven completely similarly, as for $x^{2/p} \geq 2\delta$ we derive $\dot{x} \leq -\frac{3}{4}x^{(2+p)/p}$. $\square$

Let us finally recall:

Lemma A.8   **(Gronwall's Lemma)**

Let $u : [0, T] \to \mathbb{R}$ and $a : [0, T] \to \mathbb{R}$ be continuous functions, such that $a \geq 0$. Fix $b \in \mathbb{R}$. Then,

$$u(t) \leq b + \int_0^t a(s)u(s)ds \quad \text{for all } t \in [0, T]$$

implies

$$u(t) \leq b \cdot \exp\left\{ \int_0^t a(s)ds \right\} \quad \text{for all } t \in [0, T] \,.$$

Appendix B

Bounds for SDEs

In this appendix we present technical results necessary for the proofs in Chapter 2. We establish some large deviation estimates and moment inequalities for special types of SDEs given by the amplitude equation. Estimates of these type are well known for small noise. In that case Varadhan [Var66] formulated the large deviation principle, which was used for SDEs for instance in the textbook [FW98]. For a recent publication with a nice summary of various large deviation results see [Fan03].

Nevertheless, we are not interested in the limit for noise to 0, but for the probability of large deviations for fixed noise intensity. We give some straightforward proofs for various kinds of results. Most of these results rely on the strong nonlinear dissipativity, such that we can bound exponential moments of the solutions. At the end of this section we present bounds on negative moments, which in turn immediately imply bounds for small ball probabilities. See also [Mao97] for some examples of moment inequalities for SDEs.

B.1 Large Deviation Estimate

Assumption B.4 *Let $f : \mathbb{R}^n \to \mathbb{R}^n$ and $\sigma : \mathbb{R}^n \to \mathbb{R}^n$ be locally Lipschitz functions, such that for some constants $C > 0$ and $c > 0$ we have*

$$f(x) \cdot x \leq C - c|x|^4 \quad \text{for all } x \in \mathbb{R}^n \tag{B.1}$$

and

$$|\sigma(x)| \leq C(|x| + 1) \quad \text{for all } x \in \mathbb{R}^n .$$

Consider the following stochastic Itô-differential equation.

$$dx = f(x)dt + \sigma(x)d\beta, \quad x(0) = x_0, \tag{B.2}$$

where β is a standard real-valued Brownian motion.

Remark B.6   *Of course we can consider everything in more generality. For example we could use $\sigma \in \mathbb{R}^{n \times m}$ and let β be a Brownian motion in $\mathbb{R}^m$. Nevertheless, here we stick to this simpler setting, which is sufficient for our results.*

Theorem B.9   *Under Assumption B.4. For $T > 0$, $\delta > 0$, and sufficiently small $\gamma > 0$ there is a constant $C > 0$ such that*

$$\mathbb{P}\left(\sup_{t \in [0,T]} |x(t)| \geq \eta \right) \leq Ce^{-\eta^2 \gamma} + \mathbb{P}(|x(0)| > \delta) \tag{B.3}$$

for all $\eta > 0$ and all solutions x of (B.2).

Let us first remark that with our method of proof, we will not get a better result, as we can only bound second order exponentials. Note furthermore, that in (B.4) we will see that γ is essentially bounded by c from (B.1).

The following proof is only formal, as we do not establish the existence of the moments. To proceed rigorously, we have to use suitable smooth approximations with compact support of the function $\xi \mapsto e^{\gamma \xi}$. As an example we use this rigorous approximation technique in Lemma B.11.

Proof.   Suppose first $|x(0)| \leq \delta$ almost surely. In this case we will use Itô's formula to bound exponential moments. For sufficiently small $\gamma > 0$.

$$de^{\gamma |x|^2} = 2\gamma e^{\gamma |x|^2}[x \cdot f(x)dt + x \cdot \sigma(x)d\beta + \frac{1}{2}|\sigma(x)|^2 dt + \gamma |x \cdot \sigma(x)|^2 dt]$$

$$\leq 2\gamma e^{\gamma |x|^2}[(C - c|x|^4)dt + x \cdot \sigma(x)d\beta] \tag{B.4}$$

for some constants $c, C > 0$. Hence, integrating, taking expectations, and using Gronwall's inequality yields

$$\sup_{t \in [0,T]} \mathbb{E}e^{\gamma |x|^2} \leq C \quad \text{and} \quad \mathbb{E} \int_0^T e^{\gamma |x|^2}|x|^4 dt \leq C . \tag{B.5}$$

From a comparison argument, we could see that the constant in the left equation of (B.5) is independent of T. The constant of the other equation will grow linearly in T.

Using (B.4) for $\gamma/2$ instead of γ yields

$$e^{\gamma |x(t)|^2/2} \leq e^{\gamma \delta^2/2} + C \int_0^t e^{\gamma |x(s)|^2/2}ds + \gamma \int_0^t e^{\gamma |x(s)|^2/2}x(s) \cdot \sigma(x(s))d\beta(s)$$

$$\leq C + C \int_0^T e^{\gamma |x(s)|^2/2}ds + \gamma \sup_{t \in [0,T]} \left| \int_0^t e^{\gamma |x(s)|^2/2}x(s) \cdot \sigma(x(s))d\beta(s) \right| .$$

Taking squares, using Theorem A.7 and (B.5) yields

$$\mathbb{E} \sup_{t \in [0,T]} e^{\gamma |x(t)|^2} \leq C .$$

This easily implies (B.3) for this case using Chebychev's inequality.

Suppose now that x_0 is not bounded by δ. Then define $\tilde{x}_0$ by x_0 provided $|x_0| \leq \delta$ and 0 otherwise. Let $\tilde{x}$ be the solution of (B.2) with initial condition $\tilde{x}_0$. Then $x \equiv \tilde{x}$ on the subset of the probability space, where $x_0 = \tilde{x}_0$. We derive,

$$\mathbb{P}\left(\sup_{t \in [0,T]} |x(t)| \geq \eta \right)$$

$$\leq \mathbb{P}\left(\sup_{t \in [0,T]} |\tilde{x}(t)| \geq \eta \right) + \mathbb{P}\left(|x_0| > \delta \right)$$

$$\leq C e^{-\gamma \eta^2} + \mathbb{P}\left(|x_0| > \delta \right),$$

using Chebychev's inequality and the first part of the proof. $\qquad\square$

B.2 Moment Inequalities

The following lemma provides standard a priori estimates on moments of the solution of amplitude equations like (2.15). We require weaker assumptions than the ones used in the previous section.

Assumption B.5 *Let $f : \mathbb{R}^n \to \mathbb{R}^n$ and $\sigma : \mathbb{R}^n \to \mathbb{R}^n$ be locally Lipschitz functions, such that for some constants $C > 0$ and $\delta > 0$ we have*

$$f(x) \cdot x \leq C(1 + |x|^2) \quad \text{for all } x \in \mathbb{R}^n$$

and

$$|\sigma(x)| \leq C(|x| + 1) \quad \text{for all } x \in \mathbb{R}^n .$$

Lemma B.9 *Assume Assumption B.5 is true. Given $p \geq 2, \delta > 0$, and $T_0 > 0$, there is a constant $C > 0$ such that for all solutions x of (B.2) with $\mathbb{E}|x(0)|^p \leq \delta$ we have*

$$\sup_{T \in [0,T_0]} \mathbb{E}|x(T)|^p \leq C \tag{B.6}$$

and

$$\mathbb{E} \sup_{T \in [0,T_0]} |x(T)|^p \leq C \quad \text{provided} \quad p \geq 4 .$$

If we assume a slightly stronger dissipation condition, for instance (B.1), then we can get the constant in (B.6) independent of T_0.

Proof. Again we proceed only formal, without proving that the moments exist, but this is just a matter of choosing the right approximation. Assumption B.5 and Itô's formula implies

$$d|x|^2 = 2f(x) \cdot x \, dt + 2\sigma(x) \cdot x \, d\beta + |\sigma(x)|^2 dt$$

$$\leq C(1 + |x|^2)dt + 2\sigma(x) \cdot x \, d\beta .$$

Completely analogous we derive for $p \geq 2$

$$d|x|^p \leq C(1 + |x|^p)dt + p|x|^{p-2}\sigma(x) \cdot x d\beta .$$ (B.7)

Hence,

$$\mathbb{E}|x(T)|^p \leq \mathbb{E}|x(0)|^p + CT + C \int_0^T \mathbb{E}|x(t)|^p dt .$$

Using Gronwall's inequality and the assumption on the initial condition, we finish the first part.

Furthermore (B.7) for $p/2$ instead of p yields

$$\mathbb{E} \sup_{T \in [0,T_0]} |x(T)|^p \leq C\mathbb{E}\Big(|x(0)|^{p/2} + 1 + \int_0^{T_0} |x(t)|^{p/2} dt$$

$$+ \sup_{T \in [0,T_0]} \Big| \int_0^T |x(s)|^{p-2}\sigma(x(s)) \cdot x(s) d\beta(s)| \Big)^2$$

$$\leq C + C \int_0^{T_0} \mathbb{E}|x(s)|^p ds ,$$ (B.8)

where we used the assumptions for the initial conditions and Theorem A.7. The first part now implies the second part. □

The next lemma is used for Theorem 2.6. It also provides some kind of a priori estimates for solutions of an SDE appearing in the proof of Theorem 2.6.

Lemma B.10   *Let $g(t)$ and $f(t,x)$ be stochastic processes adapted to $\{\mathcal{F}_t\}_{t \geq 0}$, the filtration of the Brownian motion β, and let $x(t)$ be a solution of*

$$dx(t) = \Big(f(t,x(t)) + g(t)\Big)dt + x(t)d\beta(t) .$$

Then for all $T_0 > 0$ and $q \geq 4$ there are some constant $C > 0$ and $c > 0$ such that for all $\delta \geq 1$ and $\gamma \geq 1$ and all solutions x with

$$\mathbb{E}|x(0)|^q \leq \delta, \quad \mathbb{E} \sup_{t \in [0,T_0]} |g(t)|^q \leq \delta, \quad \mathbb{E}|x(t)|^{q-1}|f(t,x(t))| \leq \gamma^2 \sup_{\tau \in [0,t]} \mathbb{E}|x(\tau)|^q + \delta$$

we have

$$\sup_{t \in [0,T_0]} \mathbb{E}|x(t)|^q \leq C\delta e^{c\gamma^2} \quad and \quad \mathbb{E} \sup_{t \in [0,T_0]} |x(t)|^q \leq C\delta\gamma^2 e^{c\gamma^2} .$$

Proof.   Again the proof is only formal, as we do not assume existence of moments. As usual we would have to use cut-off techniques to also establish the existence. Using Itô's formula for $p \geq 2$

$$d|x|^p = p\Big(|x|^{p-2}\langle x, f(x) + g\rangle + (p-1)|x|^p\Big)dt + p|x|^p d\beta$$ (B.9)

Now Young's inequality for $p = q$ yields

$$\mathbb{E}|x(t)|^q \leq C\delta + C \int_0^t \left(\mathbb{E}|x(\tau)|^{q-1}|f(\tau, x(\tau))| + \mathbb{E}|x(\tau)|^q\right) d\tau$$

$$\leq C\delta + C\gamma^2 \int_0^t \sup_{\tau \in [0,s]} \mathbb{E}|x(\tau)|^q ds . \tag{B.10}$$

Hence,

$$\sup_{\tau \in [0,t]} \mathbb{E}|x(\tau)|^q \leq C\delta + C\gamma^2 \int_0^t \sup_{\tau \in [0,s]} \mathbb{E}|x(\tau)|^q ds ,$$

and Gronwall's inequality yields the first inequality.

Using (B.9) with $p = q/2$ we easily derive via Young inequality and Theorem A.7 that

$$\mathbb{E} \sup_{\tau \in [0,T_0]} |x(\tau)|^q \leq C\delta + C\gamma^2 \int_0^{T_0} \sup_{\tau \in [0,s]} \mathbb{E}|x(\tau)|^q ds + C \int_0^{T_0} \mathbb{E}|x(\tau)|^q d\tau$$

and the second inequality follows from the first one. $\qquad\square$

B.2.1 *Negative Moments*

This section gives bounds for negative moments of $x(t)$. It is used in the application on pattern formation to show that the probability of x being small can be bounded from above.

Lemma B.11 *Let $x(t) \in \mathbb{R}^n$ be a solution of*

$$dx = f(x)dt + xd\beta$$

and suppose that f is locally Lipschitz with polynomial bound

$$|f(x)| \leq C|x|(1 + |x|^k)$$

for some $k > 0$. Suppose $\mathbb{E}|x(0)|^{-2p} \leq \delta$ and $\int_0^T \mathbb{E}|x(s)|^{\max\{0, k-2p\}} ds \leq \delta$ for some $p > 0$, $\delta > 0$ and $T > 0$. Then there is a constant $C > 0$ independent of δ such that

$$\sup_{t \in [0,T]} \mathbb{E}|x(t)|^{-2p} \leq C\delta .$$

If additionally $\mathbb{E}|x(0)|^{-4p} \leq \delta^2$, then

$$\mathbb{E} \sup_{t \in [0,T]} |x(t)|^{-2p} \leq C\delta .$$

Remark B.7 *We cannot expect the constants in Lemma B.11 to be independent of T, as in general $x(t) \to 0$ for $t \to \infty$ is possible, which could lead to a blow up of the negative moments. Especially, if the linear part of f is sufficiently stable.*

There is a large body of literature on path-wise and moment stability of these type of equations. But, in order to get uniform bounds, we need precise conditions on the stability of f. To be more precise, after applying Itô's formula, we need that all terms have the right sign.

Proof.   Using Itô's formula, we easily derive for all bounded $\varphi \in C^2(\mathbb{R})$, that

$$\mathbb{E}\varphi(|x(t)|^2) = \mathbb{E}\varphi(|x(0)|^2) + \int_0^t \Big(\mathbb{E}\varphi'(|x(s)|^2)\langle x(s), f(x(s))\rangle \tag{B.11}$$
$$+ \frac{1}{2}\mathbb{E}\varphi'(|x(s)|^2)|x(s)|^2 + \frac{1}{2}\mathbb{E}\varphi''(|x(s)|^2)|x(s)|^4 \Big) ds .$$

Given $p > 0$ and $\varepsilon > 0$ define $\varphi_\varepsilon(\xi) = 1/(\xi^p + \varepsilon)$ for all $\xi \geq 0$. Obviously, φ_ε is smooth and bounded with

$$\varphi_\varepsilon'(\xi) = -p\frac{\xi^{p-1}}{(\xi^p + \varepsilon)^2} \quad \text{and} \quad \varphi_\varepsilon''(\xi) = -p\frac{\xi^{p-2}(\xi^p(p-3) + \varepsilon(p-1))}{(\xi^p + \varepsilon)^3} .$$

It is easy to check that

$$|x|^{k+2p}\varphi_\varepsilon(|x|^2)^2 \leq \varphi_\varepsilon(|x|^2) + |x|^{\max\{0,k-2p\}} .$$

Thus, using the assumptions on f

$$|\varphi_\varepsilon'(|x|^2)\langle x, f(x)\rangle| \leq C\frac{|x|^{2p}}{(|x|^{2p} + \varepsilon)^2}(1 + |x|^k)$$
$$\leq C\Big(|x|^{\max\{0,k-2p\}} + \varphi_\varepsilon(|x|^2)\Big) .$$

Furthermore,

$$|\varphi_\varepsilon'(|x|^2)||x|^2 \leq p\varphi_\varepsilon(|x|^2)$$

and

$$|\varphi_\varepsilon''(|x|^2)||x|^4 \leq C\frac{|x|^{4p} + \varepsilon|x|^{2p}}{(|x|^{2p} + \varepsilon)^3} \leq C\varphi_\varepsilon(|x|^2).$$

Using that φ_ε is uniformly bounded by $\frac{1}{\varepsilon}$, it is easy to check that $\mathbb{E}\varphi_\varepsilon(|x(t)|^2)$ is differentiable with derivative in $L^\infty([0,T])$. Hence,

$$\mathbb{E}\varphi_\varepsilon(|x(t)|^2) \leq \mathbb{E}\varphi_\varepsilon(|x(0)|^2) + C\int_0^t \mathbb{E}\varphi_\varepsilon(|x(s)|^2)ds + C\int_0^T \mathbb{E}|x(s)|^{\max\{0,k-2p\}}ds .$$

Using Gronwall's inequality we derive for $t \in [0,T]$

$$\mathbb{E}\varphi_\varepsilon(|x(t)|^2) \leq C\Big(\mathbb{E}\varphi_\varepsilon(|x(0)|^2) + \delta\Big)e^{CT} \leq C\delta . \tag{B.12}$$

The monotone convergence of $\varphi_\varepsilon(\xi)$ to $|\xi|^{-p}$ implies the first claim.

For the second part Itô's formula yields for $t \in [0, T]$

$$\varphi_\varepsilon(|x(t)|^2) \le \varphi_\varepsilon(|x(0)|^2) + C \int_0^T \varphi_\varepsilon(|x(s)|^2)ds + C \int_0^T |x(s)|^{\max\{0, k-2p\}} ds$$

$$+ \sup_{t \in [0,T]} \int_0^t \varphi_\varepsilon'(|x(s)|^2)|x(s)|^2 d\beta(s) . \tag{B.13}$$

From (B.12) and Lemma A.3 we derive

$$\mathbb{E} \sup_{t \in [0,T]} \varphi_\varepsilon(|x(t)|^2) \le C\delta + C \sup_{t \in [0,T]} \left(\mathbb{E} \left(\varphi_\varepsilon(|x(t)|^2)^2 \right) \right)^{\frac{1}{2}} , \tag{B.14}$$

which in turn is bounded by $C\delta$, as $\sup_{t \in [0,T]} \mathbb{E}|x(t)|^{-4p} \le C\delta^2$, which is true by the first part. Using again monotone convergence, the claim follows. $\qquad\square$

Bibliography

[ACW83]   L. Arnold, H. Crauel, and V. Wihstutz. Stabilization of linear systems by noise. *SIAM J. Control Optim.*, 21(3):451–461, 1983.

[Arn98]   L. Arnold. *Random dynamical systems.* Springer Monographs in Mathematics. Springer-Verlag, Berlin, 1998.

[BCF93]   Z. Brzeźniak, M. Capiński, and F. Flandoli. Pathwise global attractors for stationary random dynamical systems. *Probab. Theory Related Fields*, 95(1):87–102, 1993.

[BDP05]   G. Bellettini, A. DeMasi, and E. Presutti. Tunnelling in nonlocal evolution equations. *J. Nonlinear Math. Phys.*, 12(1):50–63, 2005.

[BFR06]   D. Blömker, F. Flandoli, and M. Romito. Markovianity and ergodicity for a surface growth PDE. Preprint, arXiv.org:math.PR/0611021, 2006.

[BG02a]   N. Berglund and B. Gentz. Pathwise description of dynamic pitchfork bifurcations with additive noise. *Probab. Theory Related Fields*, 122(3):341–388, 2002.

[BG02b]   D. Blömker and C. Gugg. On the existence of solutions for amorphous molecular beam epitaxy. *Nonlinear Anal. Real World Appl.*, 3(1):61–73, 2002.

[BG03]   N. Berglund and B. Gentz. Geometric singular perturbation theory for stochastic differential equations. *J. Differential Equations*, 191(1):1–54, 2003.

[BG04]   D. Blömker and C. Gugg. Thin-film-growth-models: On local solutions. In S. Albeverio, Z.M. Ma, and M. Röckner, editors, *Recent developments in stochastic analysis and related topics*, pages 66–77. World Scientific, Singapore, 2004.

[BG06]   N. Berglund and B. Gentz. *Noise-induced phenomena in slow-fast dynamical systems. A sample-paths approach.* Probability and its Applications. Springer-Verlag, London, 2006.

[BGR02]   D. Blömker, C. Gugg, and M. Raible. Thin-film-growth models: Roughness and correlation functions. *European J. Appl. Math.*, 13(4):385–402, 2002.

[BGW07]   D. Blömker, B.. Gawron, and T. Wanner. Nucleation in the stochastic Cahn-Hilliard model – Large deviation results. In Preparation, 2007.

[BH04]   D. Blömker and M. Hairer. Multiscale expansion of invariant measures for SPDEs. *Comm. Math. Phys.*, 251(3):515 – 555, 2004.

[BH05]   D. Blömker and M. Hairer. Amplitude equations for SPDEs: Approximate centre manifolds and invariant measures. In J. Duan and E. Waymire, editors, *Probability and Partial Differential Equations in Modern Applied Mathematics*, volume 140 of *IMA Volumes in Mathematics and its Applications*, pages 41–60. Springer Verlag, 2005.

[BHP05]   D. Blömker, M. Hairer, and G.A. Pavliotis. Modulation equations: Stochastic bifurcation in large domains. *Commun. Math. Phys.*, 258(2):479–512, 2005.

[BHP06]   D. Blömker, M. Hairer, and G.A. Pavliotis. Multiscale analysis for SPDEs with quadratic nonlinearities. Preprint, arXiv.org:math.PR/0611537, 2006.

[BHP07]   D. Blömker, M. Hairer, and G.A. Pavliotis. Stochastic Swift-Hohenberg equation near a change of stability. To appear in Proceedings of the Equadiff 2005, 2007.

[Blö03]   D. Blömker. Amplitude equations for locally cubic non-autonomous nonlinearities. *SIAM Journal on Applied Dynamical Systems*, 2(2):464–486, 2003.

[Blö05a]   D. Blömker. Approximation of the stochastic Rayleigh-Bénard problem near the onset of instability and related problems. *Stoch. Dyn.*, 5(3):441–474, 2005.

[Blö05b]   D. Blömker. Non-homogeneous noise and Q-Wiener processes on bounded domains. *Stochastic Anal. Appl.*, 23(2):255–273, 2005.

[BMPS01]   D. Blömker, S. Maier-Paape, and G. Schneider. The stochastic Landau equation as an amplitude equation. *Discrete Contin. Dyn. Syst. Ser. B*, 1(4):527–541, 2001.

[BMPW01]   D. Blömker, S. Maier-Paape, and T. Wanner. Spinodal decomposition for the Cahn-Hilliard-Cook equation. *Commun. Math. Phys.*, 223(3):553–582, 2001.

[BMPW05]   D. Blömker, S. Maier-Paape, and T. Wanner. Phase separation in stochastic Cahn-Hilliard models. In A. Miranville, editor, *Mathematical Methods and Models in Phase Transitions*, pages 1–41. Nova Science Publisher, New York, 2005.

[BMPW07]   D. Blömker, S. Maier-Paape, and T. Wanner. Second phase spinodal decomposition for the Cahn-Hilliard-Cook equation. *Trans. Am. Math. Soc.*, to appear., 2007.

[BP99]   Z. Brzeźniak and S. Peszat. Space-time continuous solutions to SPDE's driven by a homogeneous Wiener process. *Studia Math.*, 137(3):261–299, 1999.

[BP00]   Z. Brzeźniak and S. Peszat. Strong local and global solutions for stochastic Navier-Stokes equations. In *Infinite dimensional stochastic analysis (Amsterdam, 1999)*, volume 52 of *Verh. Afd. Natuurkd. 1. Reeks. K. Ned. Akad. Wet.*, pages 85–98. R. Neth. Acad. Arts Sci., Amsterdam, 2000.

[Bra91]   S. Brassesco. Some results on small random perturbations of an infinite-dimensional dynamical system. *Stochastic Process. Appl.*, 38(1):33–53, 1991.

[BS95]   A.-L. Barabási and H.E. Stanley. *Fractal concepts in surface growth*. Cambridge University Press, Cambridge, 1995.

[CB95]   Z.R. Cuerno and A.-L. Barabási. Dynamic scaling of ion-sputtered surfaces. *Phys. Rev. Lett.*, 74:4746–4749, 1995.

[CDF97]   H. Crauel, A. Debussche, and F. Flandoli. Random attractors. *J. Dynam. Differential Equations*, 9(2):307–341, 1997.

[CE90]   P. Collet and J.-P. Eckmann. The time dependent amplitude equation for the Swift-Hohenberg problem. *Comm. Math. Phys.*, 132(1):139–153, 1990.

[Cer01]   S. Cerrai. *Second order PDE's in finite and infinite dimension. A probabilistic approach.*, volume 1762 of *Lecture Notes in Mathematics*. Springer-Verlag, Berlin, 2001.

[Cer05]   S. Cerrai. Stabilization by noise for a class of stochastic reaction-diffusion equations. *Probab. Theory Relat. Fields*, 133(2):190–214, 2005.

[CF94]   H. Crauel and F. Flandoli. Attractors for random dynamical systems. *Probab. Theory Related Fields*, 100(3):365–393, 1994.

[CF98]   H. Crauel and F. Flandoli. Additive noise destroys a pitchfork bifurcation. *J. Dynam. Differential Equations*, 10(2):259–274, 1998.

[CH93]   M.C. Cross and P.C. Hohenberg. Pattern formation outside of equilibrium. *Rev. Mod. Phys.*, 65:851–1112, 1993.

[Chu02]   I. Chueshov. *Monotone random systems theory and applications*, volume 1779 of *Lecture Notes in Mathematics*. Springer-Verlag, Berlin, 2002.

[CIS99]   H. Crauel, P. Imkeller, and M. Steinkamp. Bifurcations of one-dimensional stochastic differential equations. In *Stochastic dynamics (Bremen, 1997)*, pages 27–47. Springer, New York, 1999.

[CKS04]   T. Caraballo, P.E. Kloeden, and B. Schmalfuß. Exponentially stable stationary solutions for stochastic evolution equations and their perturbation. *Appl. Math. Optim.*, 50(3):183–207, 2004.

[CLR00]   T. Caraballo, J.A. Langa, and J.C. Robinson. Stability and random attractors for a reaction-diffusion equation with multiplicative noise. *Discrete Contin. Dynam. Systems*, 6(4):875–892, 2000.

[CLR01]   T. Caraballo, J.A. Langa, and J.C. Robinson. A stochastic pitchfork bifurcation in a reaction-diffusion equation. *R. Soc. Lond. Proc. Ser. A Math. Phys. Eng. Sci.*, 457(2013):2041–2061, 2001.

[CR04]   T. Caraballo and J.C. Robinson. Stabilisation of linear PDEs by Stratonovich noise. *Systems Control Lett.*, 53(1):41–50, 2004.

[Cra01]   H. Crauel. Random point attractors versus random set attractors. *J. London Math. Soc. (2)*, 63(2):413–427, 2001.

[CS04]   I. Chueshov and M. Scheutzow. On the structure of attractors and invariant measures for a class of monotone random systems. *Dynamical Systems: An International Journal*, 19(2):127–144, 2004.

[DE00]   J. Duan and V.J. Ervin. On nonlinear amplitude evolution under stochastic forcing. *Appl. Math. Comput.*, 109(1):59–65, 2000.

[DE01]   J. Duan and V.J. Ervin. On the stochastic Kuramoto-Sivashinsky equation. *Nonlinear Anal.*, 44(2, Ser. A: Theory Methods), 2001.

[DEKS95]   A. Doelman, W. Eckhaus, R. Kuske, and R. Schielen. Pattern formation in systems on spatially periodic domains. In *Nonlinear dynamics and pattern formation in the natural environment (Noordwijkerhout, 1994)*, volume 335 of *Pitman Res. Notes Math. Ser.*, pages 85–105. Longman, Harlow, 1995.

[DKS01]   J. Duan, P.E. Kloeden, and B. Schmalfuß. Exponential stability of the quasigeostrophic equation under random perturbations. In *Stochastic climate models (Chorin, 1999)*, volume 49 of *Progr. Probab.*, pages 241–256. Birkhäuser, Basel, 2001.

[DLS03]   J. Duan, K. Lu, and B. Schmalfuß. Invariant manifolds for stochastic partial differential equations. *Ann. Probab.*, 31(4):2109–2135, 2003.

[DLS04]   J. Duan, K. Lu, and B. Schmalfuß. Smooth stable and unstable manifolds for stochastic evolutionary equations. *Dyn. Differ. Equations*, 16(4):949–972, 2004.

[DPDT94]   G. Da Prato, A. Debussche, and R. Temam. Stochastic Burgers' equation. *NoDEA Nonlinear Differential Equations Appl.*, 1(4):389–402, 1994.

[DPZ92]   G. Da Prato and J. Zabczyk. *Stochastic equations in infinite dimensions*, volume 44 of *Encyclopedia of Mathematics and its Applications*. Cambridge University Press, Cambridge, 1992.

[DPZ96]   G. Da Prato and J. Zabczyk. *Ergodicity for infinite-dimensional systems*, volume 229 of *London Mathematical Society Lecture Note Series*. Cambridge University Press, Cambridge, 1996.

[DPZ02]   G. Da Prato and J. Zabczyk. *Second order partial differential equations in Hilbert spaces*, volume 293 of *London Mathematical Society Lecture Note Series*. Cambridge University Press, Cambridge, 2002.

[DTZ05a] I.M. Davies, A. Truman, and H. Zhao. Stochastic heat and Burgers equations and their singularities. In E.C. Waymire and J. Duan, editors, *Probability and partial differential equations in modern applied mathematics.*, volume 140 of *The IMA Volumes in Mathematics and its Applications*, pages 79–95. Springer, New York, 2005.

[DTZ05b] I.M. Davies, A. Truman, and H. Zhao. Stochastic heat and Burgers equations and their singularities. II: Analytical properties and limiting distributions. *J. Math. Phys.*, 46(4):043515, 2005.

[DW05]   C. Diks and F. Wagener. Equivalence and bifurcations of finite order stochastic processes. Preprint, Tinbergen Institute Discussion Papers 05-043/1, Tinbergen Institute., 2005.

[DW06a]   C. Diks and F. Wagener. A weak bifurcation theory for discrete time stochastic dynamical systems. Preprint, Tinbergen Institute Discussion Papers 06-043/1, Tinbergen Institute., 2006.

[DW06b]   J. Duan and W. Wang. Invariant manifold reduction and bifurcation for stochastic partial differential equation. Preprint, oai:arXiv.org:math/0607050, 2006.

[EH01]   J.-P. Eckmann and M. Hairer. Uniqueness of the invariant measure for a stochastic PDE driven by degenerate noise. *Comm. Math. Phys.*, 219(3):523–565, 2001.

[EL02]   W. E and D. Liu. Gibbsian dynamics and invariant measures for stochastic dissipative PDEs. *J. Statist. Phys.*, 108(5-6):1125–1156, 2002. Dedicated to David Ruelle and Yasha Sinai on the occasion of their 65th birthdays.

[EM01]   W. E and J.C. Mattingly. Ergodicity for the Navier-Stokes equation with degenerate random forcing: Finite-dimensional approximation. *Comm. Pure Appl. Math.*, 54(11):1386–1402, 2001.

[Fan03]   M. Fantozzi. Large deviations for semilinear differential stochastic equations with dissipative non-linearities. *Stochastic Anal. Appl.*, 21(1):127–139, 2003.

[FBK⁺02] S. Facsko, T. Bobek, H. Kurz, T. Dekorsy, S. Kyrsta, and R. Cremer. Ion-induced formation of regular nanostructures on amorphous GaSb surfaces. *Applied Physics Letters*, 80(1):130–132, 2002.

[FS96]   F. Flandoli and B. Schmalfuß. Random attractors for the 3D stochastic Navier-Stokes equation with multiplicative white noise. *Stochastics Stochastics Rep.*, 59(1-2):21–45, 1996.

[FS99]   F. Flandoli and B. Schmalfuß. Weak solutions and attractors for three-dimensional Navier-Stokes equations with nonregular force. *J. Dynam. Differential Equations*, 11(2):355–398, 1999.

[FW98]   M. I. Freidlin and A. D. Wentzell. *Random perturbations of dynamical systems*, volume 260 of *Grundlehren der Mathematischen Wissenschaften*. Springer-Verlag, New York, second edition, 1998. Translated from the 1979 Russian original by Joseph Szücs.

[Gaw06]   B. Gawron. *Nucleation in the one-dimensional Cahn-Hilliard Model.* PhD thesis, RWTH Aachen, 2006.

[Get98]   A.V. Getling. *Rayleigh-Bénard convection. Structures and dynamics.*, volume 11 of *Advanced Series in Nonlinear Dynamics.* World Scientific Publishing Co. Inc., River Edge, NJ, 1998.

[GKS04]   D. Givon, R. Kupferman, and A. Stuart. Extracting macroscopic dynamics: model problems and algorithms. *Nonlinearity*, 17(6):R55–R127, 2004.

[GM01]   B. Goldys and B. Maslowski. Uniform exponential ergodicity of stochastic dissipative systems. *Czechoslovak Math. J.*, 51(126)(4):745–762, 2001.

[Hak83]   H. Haken. *Synergetics: An introduction. Nonequilibrium phase transitions and self-organization in physics, chemistry, and biology*, volume 1 of *Springer Series in Synergetics.* Springer-Verlag, Berlin, 3rd edition, 1983.

[Hal80]   J.K. Hale. *Ordinary differential equations.* Robert E. Krieger Publishing Co. Inc., Huntington, N.Y., 2nd edition, 1980.

[Hen81]   D. Henry. *Geometric theory of semilinear parabolic equations*, volume 840 of *Lecture Notes in Mathematics.* Springer-Verlag, Berlin, 1981.

[HHZ95]   T. Halpin-Healy and Y. Zhang. Kinetic roughening phenomena, stochastic growth, directed polymers and all that. *Physics Reports*, 354:214–414, 1995.

[HM06]   M. Hairer and J.C. Mattingly. Ergodicity of the 2D Navier-Stokes equations with degenerate stochastic forcing. *Annals of Mathematics*, 164(3):993–1032, 2006.

[HS92]   P.C. Hohenberg and J.B. Swift. Effects of additive noise at the onset of Rayleigh-Bénard convection. *Physical Review A*, 46:4773–4785, 1992.

[HS01]   E. Hausenblas and J. Seidler. A note on maximal inequality for stochastic convolutions. *Czechoslovak Math. J.*, 51(126)(4):785–790, 2001.

[HT94]   W. Hackenbroch and A. Thalmaier. *Stochastische Analysis. Eine Einführung in die Theorie der stetigen Semimartingale.* Mathematische Leitfäden. B. G. Teubner, Stuttgart, 1994.

[Kie04]   H. Kielhöfer. *Bifurcation theory. An introduction with applications to PDEs.*, volume 156 of *Applied Mathematical Sciences.* Springer-Verlag, New York, 2004.

[KPS04]   K. Kupferman, G.A. Pavliotis, and A. Stuart. Ito versus Stratonovich white noise limits for systems with inertia and colored multiplicative noise. *Phys. Rev. E*, 70:036120, 2004.

[KRZD99]   N.V. Krylov, M. Röckner, J. Zabczyk, and G. DaPrato, editors. *Stochastic PDE's and Kolmogorov equations in infinite dimensions.*, volume 1715 of *Lecture Notes in Mathematics.* Springer, Berlin, 1999.

[KS01]   S. Kuksin and A. Shirikyan. A coupling approach to randomly forced nonlinear PDE's. I. *Comm. Math. Phys.*, 221(2):351–366, 2001.

[KSM92]   P. Kirrmann, G. Schneider, and A. Mielke. The validity of modulation equations for extended systems with cubic nonlinearities. *Proc. R. Soc. Edinb., Sect. A*, 122(1-2):85–91, 1992.

[KSW03]   B.B. King, O. Stein, and M. Winkler. A fourth-order parabolic equation modeling epitaxial thin film growth. *J. Math. Anal. Appl.*, 286(2):459–490, 2003.

[Kur73]   T.G. Kurtz. A limit theorem for perturbed operator semigroups with applications to random evolutions. *J. Functional Analysis*, 12:55–67, 1973.

[Kus03]   R. Kuske. Multi-scale analysis of noise-sensitivity near a bifurcation. In *IUTAM Symposium on Nonlinear Stochastic Dynamics*, volume 110 of *Solid*

             Mech. Appl., pages 147–156. Kluwer Acad. Publ., Dordrecht, 2003.

[Kwi02]   A.A. Kwiecińska. Stabilization of evolution equations by noise. *Proc. Amer. Math. Soc.*, 130(10):3067–3074, 2002.

[LDS91]   Z.W. Lai and S. Das Sarma. Kinetic growth with surface relaxation: Continuum versus atomistic models. *Phys. Rev. Lett.*, 66:2348–2351, 1991.

[LM99]   G. Lythe and E. Moro. Dynamics of defect formation. *Physical Review E*, 59:R1303–1306, 1999.

[Lun95]   A. Lunardi. *Analytic semigroups and optimal regularity in parabolic problems*, volume 16 of *Progress in Nonlinear Differential Equations and their Applications*. Birkhäuser Verlag, Basel, 1995.

[LV05]   S.J. Linz and S. Vogel. Continuum modeling of sputter erosion under normal incidence: Interplay between nonlocality and nonlinearity. *Phys. Rev. B*, 72:035416, 2005.

[Lyt96]   G. Lythe. Domain formation in transitions with noise and a time-dependent bifurcation parameter. *Physical Review E*, 53:R4271–4274, 1996.

[Mao97]   X. Mao. *Stochastic differential equations and their applications*. Horwood Publishing Series in Mathematics & Applications. Horwood Publishing Limited, Chichester, 1997.

[Mat02]   J.C. Mattingly. Exponential convergence for the stochastically forced Navier-Stokes equations and other partially dissipative dynamics. *Comm. Math. Phys.*, 230(3):421–462, 2002.

[MC00]   P.C. Matthews and S.M. Cox. Pattern formation with a conservation law. *Nonlinearity*, 13(4):1293–1320, 2000.

[McK69]   H.P. McKean, Jr. *Stochastic integrals*, volume 5 of *Probability and Mathematical Statistics*. Academic Press, New York, 1969.

[MS95]   A. Mielke and G. Schneider. Attractors for modulation equations on unbounded domains – Existence and comparison. *Nonlinearity*, 8(5):743–768, 1995.

[MS01]   B. Maslowski and I. Simão. Long-time behaviour of nonautonomous SPDE's. *Stochastic Process. Appl.*, 95(2):285–309, 2001.

[MSZ00]   A. Mielke, G. Schneider, and A. Ziegra. Comparison of inertial manifolds and application to modulated systems. *Math. Nachr.*, 214:53–69, 2000.

[MT93]   S.P. Meyn and R.L. Tweedie. *Markov chains and stochastic stability*. Communications and Control Engineering Series. Springer-Verlag London Ltd., London, 1993.

[MTVE01] A.J. Majda, I. Timofeyev, and E. Vanden-Eijnden. A mathematical framework for stochastic climate models. *Comm. Pure Appl. Math.*, 54(8):891–974, 2001.

[MZZ07]   S.-E.A. Mohammed, T. Zhang, and H. Zhao. The stable manifold theorem for semilinear stochastic evolution equations and stochastic partial differential equations. To appear in *Mem. Am. Math. Soc.*, 2007.

[OA03]   J. Oh and G. Ahlers. Thermal-noise effect on the transition to Rayleigh-Bnard convection. *Phys. Rev. Lett.*, 91:094501, 2003.

[OdZSA04] J. Oh, J.M. Ortiz de Zrate, J.V. Sengers, and G. Ahlers. Dynamics of fluctuations in a fluid below the onset of Rayleigh-Bénard convection. *Phys. Rev. E*, 69:021106, 2004.

[OKRVE06]F. Otto, R.V. Kohn, M.G. Reznikoff, and E. Vanden-Eijnden. Action minimization and sharp-interface limits for the stochastic Allen-Cahn equation. *Comm. Pure App. Math*, 2006.

[Øks98]   B. Øksendal. *Stochastic differential equations. An introduction with applica-*

tions. Universitext. Springer-Verlag, Berlin, 5th edition, 1998.

[OV05]   E. Olivieri and M.E. Vares. *Large deviations and metastability.* Cambridge University Press, Cambridge, 2005.

[Paz83]   A. Pazy. *Semigroups of linear operators and applications to partial differential equations*, volume 44 of *Applied Mathematical Sciences*. Springer-Verlag, New York, 1983.

[PS03]   G.A. Pavliotis and A.M. Stuart. White noise limits for inertial particles in a random field. *Multiscale Model. Simul.*, 1(4):527–533 (electronic), 2003.

[Rac91]   S.T. Rachev. *Probability metrics and the stability of stochastic models.* Wiley Series in Probability and Mathematical Statistics: Applied Probability and Statistics. John Wiley & Sons Ltd., Chichester, 1991.

[Ris84]   H. Risken. *The Fokker-Planck equation. Methods of solution and applications.*, volume 18 of *Springer Series in Synergetics*. Springer-Verla, Berlin etc., 1984.

[RK06]   Y. Tonegawa R.V. Kohn, M.G. Reznikoff. Sharp-interface limit of the Allen-Cahn action functional in one space dimension. *Calc. Var. Partial Differential Equations*, 25(4):503–534, 2006.

[RLH00]   M. Raible, S.J. Linz, and P. Hänggi. Amorphous thin film growth: Minimal deposition equation. *Physical Review E*, 62:1691–1705, 2000.

[RML$^+$00]   M. Raible, S.G. Mayr, S.J. Linz, M. Moske, P. Hänggi, and K. Samwer. Amorphous thin film growth: Theory compared with experiment. *Europhysics Letters*, 50:61–67, 2000.

[Rob02]   J.C. Robinson. Stability of random attractors under perturbation and approximation. *J. Differential Equations*, 186(2):652–669, 2002.

[Rob03]   A.J. Roberts. A step towards holistic discretisation of stochastc partial differential equations. *ANZIAM J.*, 45(E):C1–C15, 2003.

[RS04]   M. Röckner and Z. Sobol. A new approach to Kolmogorov equations in infinite dimensions and applications to stochastic generalized Burgers equations. *C. R. Acad. Sci. Paris, Serie I*, 338(12):945–949, 2004.

[RS06]   M. Röckner and Z. Sobol. Kolmogorov equations in infinite dimensions: Well-posedness, regularity of solutions, and applications to stochastic generalized Burgers equations. *Ann. Prob*, 34(2):663–727, 2006.

[SA02]   M.A. Scherer and G. Ahlers. Temporal and spatial properties of fluctuations below a supercritical primary bifurcation to traveling oblique-roll electroconvection. *Phys. Rev. E*, 65:051101, 2002.

[Sch94]   G. Schneider. Error estimates for the Ginzburg-Landau approximation. *Z. Angew. Math. Phys.*, 45(3):433–457, 1994.

[Sch96]   G. Schneider. The validity of generalized Ginzburg-Landau equations. *Math. Methods Appl. Sci.*, 19(9):717–736, 1996.

[Sch97]   B. Schmalfuß. The random attractor of the stochastic Lorenz system. *Z. Angew. Math. Phys.*, 48(6):951–975, 1997.

[Sch98]   B. Schmalfuß. A random fixed point theorem and the random graph transformation. *J. Math. Anal. Appl.*, 225(1):91–113, 1998.

[Sch99]   B. Schmalfuß. Measure attractors and random attractors for stochastic partial differential equations. *Stochastic Anal. Appl.*, 17(6):1075–1101, 1999.

[Sch01]   G. Schneider. Bifurcation theory for dissipative systems on unbounded cylindrical domains. – An introduction to the mathematical theory of modulation equations. *ZAMM, Z. Angew. Math. Mech.*, 81(8):507–522, 2001.

[SH77]   J. Swift and P.C. Hohenberg. Hydrodynamic fluctuations at the convective instability. *Phys. Rev. A*, 15(1):319–328, 1977.

[SP94]   M. Siegert and M. Plischke. Solid-on-solid models of molecular-beam epitaxy. *Physical Review E*, 50:917–931, 1994.

[SR94]   W. Schöpf and I. Rehberg. The influence of thermal noise on the onset of travelling-wave convection in binary fluid mixtures: An experimental investigation. *J. Fluid Mech.*, 271:235–265, 1994.

[Ste00] M. Steinkamp. *Bifurcations of one dimensional stochastic differential equations*. PhD thesis, Berlin: Logos Verlag. Berlin: Humboldt-Univ. Berlin, Mathematisch-Naturwissenschaftliche Fakultät II, 194 p., 2000.

[SW05]   O. Stein and M. Winkler. Amorphous molecular beam epitaxy: Global solutions and absorbing sets. *European Journal of Applied Mathematics*, 16:767–798, 2005.

[Tea06a] O. Tearne. *Collapse of Random Attractors for Dissipative SDEs*. PhD thesis, University of Warwick, 2006.

[Tea06b] O.M. Tearne. Collapse of attractors for ODEs under small random perturbations. Preprint, 2006.

[Tub84] L. Tubaro. An estimate of Burkholder type for stochastic processes defined by the stochastic integral. *Stochastic Anal. Appl.*, 2(2):187–192, 1984.

[Var66] S.R.S. Varadhan. Asymptotic probabilities and differential equations. *Comm. Pure Appl. Math.*, 19:261–286, 1966.

[Wal86] J.B. Walsh. An introduction to stochastic partial differential equations. In *École d'été de probabilités de Saint-Flour, XIV—1984*, volume 1180 of *Lecture Notes in Math.*, pages 265–439. Springer, Berlin, 1986.

[Wal97] D. Walgraef. *Spatio-Temporal Pattern Formation*. Partially ordered systems. Springer, 1997.

[Wei80] J. Weidmann. *Linear operators in Hilbert spaces*, volume 68 of *Graduate Texts in Mathematics*. Springer-Verlag, New York, 1980. Translated from the German by Joseph Szücs.

[WSPS04] J. Walter, C. Schütte, G.A. Pavliotis, and A.M. Stuart. Averaging of stochastic differential equations: Kurtz's theorem revisited. Preprint, 2004.

[Zäh98] M. Zähle. Integration with respect to fractal functions and stochastic calculus. I. *Probab. Theory Related Fields*, 111(3):333–374, 1998.

Index

A, linear operator, 30, 43, 56
$AC([0,T],\mathbb{R})$, 106
B, bilinear operator, 43
L, linear operator, 29, 56
$N(L)$, kernel of L, 29
P_c, projection onto $\mathcal{N}$, 29
$P_s = I - P_c$, 29
Q-Wiener process, 57
  trace-class, 62
Q_ε, covariance operator, 95
$R = u - \varepsilon w$, remainder, 34
$T = \varepsilon^2 t$, slow time-scale, 9
W_L, stochastic convolution, 17, 57
$W_{\mathcal{L}_\varepsilon}$, stochastic convolution, 93
X, Hilbert space, 29
X^α, fractional space, 29
$\Re(z)$, real part, 15
$\mathcal{F}$, trilinear operator, 30, 56
$\mathcal{L}(X,Y)$, continuous linear, 29
$\mathcal{L} := -P(i\partial_x)$, 97
$\mathcal{L}_\varepsilon$, differential operator, 92
$\mathcal{L}_k(X,Y)$, continuous k-linear, 29
$\mathcal{N} := N(L)$, kernel, 29
$\mathcal{O}$-notation, 19
$\mathcal{P}_T$, Markov semigroup, 73
$\mathcal{Q}_T$, Markov semigroup, 73
$\mathcal{S} = \mathcal{N}^\perp$, 56
$\mathcal{T}_\varepsilon(T)$, bad times, 77
c.c., complex conjugate, 15
$d\beta$, Itô differential, 30
$e_k(x) = e^{ik\pi x/L}/\sqrt{2L}$, 98
q, q_ε, correlation function, 91
$v = u - W_L$, 59
e^{tL}, analytic semigroup, 29

absolutely continuous, 106

admissible random variables, 100
amplitude equation, 9, 33, 63
analytic semigroup e^{tL}, 29
approximation, 21
  invariant measure, 72
  large domain, 101
    invariant measure, 102
    stochastic convolution, 95, 99
  random attractor, 87
  theorem, 35, 47, 65
    global, 61
    local, 61
approximative centre manifold, 80
attractivity, 19
  large domain, 101
  local, 19
  theorem, 32, 45
    global, 60
    local, 59, 63

Burgers' equation, 11
Burkholder-Davis-Gundy, 104

comparison argument, 106

Delta-distribution, δ, 43
Doob's theorem, 104
Dual Markov semigroup, 70

factorisation method, 105
fast modes, 18
fractional space X^α, 29

Ginzburg-Landau equation, 21
  complex, 92

invariant measure, 70
  approximation
    idea, 70
    theorem, 72
    total variation, 74
  stationary realization, 70
Itô differential, $d\beta$, 30

Kantorovich distance, 71
Kuramoto-Sivashinsky equation, 23

large domains, 13, 90

maximal inequalities, 104
modulated pattern, 90

noise, 7
  additive, 56
  complex, 92
  large domains, 91
  long-range correlations, 95
  multiplicative, 26
  non-degenerate, 70
  space-time white, 7
  trace-class, 62
nonlinear dissipativity, 30
nonlinear stability condition, 30
nonlinearity
  cubic, 8, 30, 56
  other, 10
  quadratic, 11, 43

pattern space $\mathcal{N}$, 76
pull-back convergence, 84

random
  attractor, 85
  dynamical system, 84
  fixed points, 82
  invariant manifolds, 80
Rayleigh-Bénard problem, 23
residual
  definition, 20, 64
  theorem, 21, 33, 45, 64

slow modes, 18
slow time-scale $T = \varepsilon^2 t$, 9
solution
  mild, 31, 44, 58
    global, 59
    local, 62
  strong, 31
stationary solution, 70
stochastic convolution
  large domain $W_{\mathcal{L}_\varepsilon}$, 93
stochastic convolution W_L, 57
stochastic convolution $W_L(t)$, 17
surface growth equation, 22
Swift-Hohenberg equation, 8, 21, 75
  large domains, 91

total variation norm, 71

variation of constants formula, 17, 31, 44

Wasserstein norm, 71
Wiener process, 57
  trace-class, 62